IEE CONTROL ENGINEERING SERIES

SERIES EDITORS: G.A.MONTGOMERIE
                PROF. H.NICHOLSON

Control in Hazardous Environments

R.E. Young

Previous volumes in this series:

Control in Hazardous Environments

R.E. Young

R.E. YOUNG, B.Sc.(Eng.), M.R.Ae.S., C.Eng., F.I.E.E.

PETER PEREGRINUS LTD.
on behalf of the
Institution of Electrical Engineers

Published by: Peter Peregrinus Ltd.,Stevenage, and New York
© 1982: Peter Peregrinus Ltd.

British Library Cataloguing in Publication Data

Young, R.E.
    Control in hazardous environments - (IEE control
    engineering series: 17)
    1.   Control engineering
    I.   Title   II.   Series
    629.8   TJ213

ISBN: 0-906048 69 9

Printed in England by Short Run Press Ltd.,Exeter

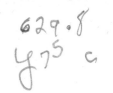

CONTENTS

PREFACE

PREFACE

'Control in Hazardous Environments' owes much to contacts
established within e.g. public utilities and manufacturing
industry in a number of countries including Canada and
Australia; and which have given perspective obtainable in no
other way.

Concerned essentially with the safe operation of centrally
controlled 'complexes' such as oil and nuclear power installations, this book is intended to serve Engineers and technolog-
ically-based administrators in Government and Industry.

It has also proved possible to take advantage of relevant
aerospace practice in many of the proposals that are put
forward particuarly for environmental protection and in connection with full 'Crisis Management'. As indicated in this book,
'blanket' applications of aerospace techniques and attitudes in
other fields are usually not as successful as might be expected.

However, from experience it can be stated that during
development there is a relatively short phase where methods and
techniques are not firmly committed and can be transferred
satisfactorily to other areas of application; and in the book
it will be found that references to earlier work have been made
in these terms, i.e. where the design principles are clear but
have not been brought to a specialist level.

For students it is felt that the greatest value of the book
will be to give a new concept of High Technology Systems
Engineering, not only in the context of Safety, but also for the
link-up made between design in its widest sense and Research and
Development as generally understood.

In conclusion the author would like to express his
appreciation of discussions with Professor Harold C.A. Hankins
of the University of Manchester Institute of Science and
Technology particularly with regard to certain psycho-physiological aspects involved here; and to acknowledge the

contribution to Mr. John A.W. Robson of the British Petroleum
Company Limited to the specialist background of this book. A
personal acknowledgement is also made to Mr. Stuart Poole of
Manchester Polytechnic for his protracted development testing and
evaluation of transducers.

R.E. YOUNG

Chapter 1

INTRODUCTION AND REVIEW

1.1 'Known hazards' and 'Incident' situations - Crisis Management

By their very nature, centrally controlled 'complexes' -
e.g. oil, chemical and nuclear power installations - bring
certain safety hazards with them.
These fall into two basic divisions :
(a) 'Known' hazards exemplified by the risk of explosion in
coal mines or at oil well-heads. The general design of such
installations is well established; and particularly in the
context of Safety, comprehensive Codes of Practice have been
evolved and followed over the years.
These direct safety considerations apply, of course, both
to the overall control system and to the main system (plant)
which it serves; and this implied 'Total System' design approach
is ideally carried out as a coordinated exchange between user
and control system designer.(Ref.1)
(b) In contrast, the second division of hazards are associated
with 'Incident' conditions, effectively states of emergency,
which are set up when a completely unforeseen - unexpected -
failure develops. Such situations can, in extreme cases, reach
'crisis' proportions.

Crisis Management - Divisions

Because of their unpredictable character, 'incident' failure
threatening the whole operation of the plant must be countered
by human intervention i.e. an operator must take over-riding
control.
Thus although precautions can be taken in design to counter

such fault conditions (see next Section), it is virtually
impossible to make provision for all the combinations of faults
which might be encountered, not least in cases where they develop
as a 'consequential' ('knock-on') build-up.

Therefore there is an evident requirement for the special
operational facilities of 'Crisis Management' to be made
available for control systems whenever it is adjudged that the
probability of a destructive accident may exist. This
requirement has, of course, been highlighted during recent years
by the near-catastrophic accidents which, though isolated, have
been experienced in many industrialised countries in the world.

The whole question of Crisis Management is treated in some
detail in Chapter 5 - Overall System Design, and, in specific
contexts, in earlier chapters; but at this point it may be
stated that it falls into two main functional divisions. The
first of these, Crisis Control i.e. Crisis Management proper, may
be defined as an intervention action where an operator takes
over-riding control in the event of an emergency, first to
contain it and then literally to bring it under control.

The second of these functional divisions embraces techniques
aimed at forestalling as far as possible the development of
Incident conditions. In its most effective form, such
'Anticipatory Control' is aimed at picking up the first signs
of a single failure which, with conventional methods would
pass undetected, and which, as a result would have just as
disastrous consequences as a complete 'knock-on' build-up. Taking
up an intermediate position is the provision of a more centralised 'over-view' capability, designed to give early warning
(information) of faults during the onset period preceding the
development of a full Incident situation.

Crisis Management - Basic system approach

To enable all forms of Crisis Management to be carried out
successfully a main technical requirement is for the Control
Engineer to have a clear and accurate picture of the operation
of the system always available to him. This brings in the need

for maximum 'protection' to be built in to the system. The types
of protection should include the 'instantaneous' presentation of
critical data - not least to cover the rapidly changing events
of a major incident; while, for example, provisions should be
made to ensure that all control information is as highly 'secure'
as possible.

Also, in this general connection, selective measurement and
'state' information must be presented to the Control Engineer in
such a way that they are assimilated almost sub-consciously so
that delay and error, particularly ambiguity, are kept to an
absolute minimum. The whole question of the two-way transfer of
information at the 'man-machine' interface in the control room is
covered broadly in Chapter 4, where this is extended to include
some of the psycho-physiological aspects that should be taken
into account when designing for Crisis Management; that is, where
analysis and decision-making under some degree of stress are
demanded of the operator.

An analogy can be drawn with Aircraft (piloting) Control for
this rapidly changing operational conditions of ultimate Incident
control. There is one example of the larger parallel which can
be seen with 'Aerospace' methods and practice in this whole field
and which is brought in at appropriate parts of this book.

It must be noted that 'blanket' applications of Aerospace
techniques and attitudes to 'Public Utility Telemetry' and
Central (Supervisory) Control generally are not as successful as
might be expected at first sight. This is due to a variety of
reasons, of which one of the most telling is the need for a
specialised approach to each field of application as implied by
the 'Total System' design requirement of 1.1(a). Moreover with
the 'slow-to-change' parameters of most established public utility
and like, systems, it has proved possible to keep to accepted
essentially routine control system design and - from long
experience - to cover most of the likely fault conditions. The
implications of this policy in relation to the more complex
computer-based systems are discussed in the next main Section.

However, with the continuing growth in size of modern

centrally controlled installations and the corresponding increase
in vulnerability with regard to Safety, it becomes clear that the
point is being reached where provision must be made to deal with
every form of emergency - however unlikely - that can be envisaged, and on an 'immediate' basis.

This requirement has been reinforced by the actual
occurrence of near-catastrophic accidents as mentioned earlier;
and steps are being taken internationally to develop more
effective safety precautions from the base of existing Rules and
Codes of Practice. Examples are the preliminary announcement of
a new EEC Directive aimed at avoiding major industrial disasters
(UK newspaper report July 1 1980); and an international Atomic
Energy Agency conference called to harmonise nuclear reactor
safety standards worldwide (UK newspaper report October 20 1980).

Also, in the context of the prevention of destructive
accidents, a major contribution has been made in the United
Kingdom by the Advisory Committee on Major Hazards in their
second report (Ref.2) prepared under the aegis of the Health and
Safety Commission

The new safety regime in control

It becomes evident that as soon as it is decided that full
Crisis Management facilities must be incorporated in an
installation, then an entirely new form of 'Control Working' -
a new 'Safety Regime' - must be set up. This approach may be
summarised as operation in 'real time' i.e. where, as already
described, the presentation of critical information to the
Control Engineer must be 'immediate', with maximum 'security'
being given to this information.

Taking the first aspect of this new regime, for the presentation of information to be effective, techniques covered by the
term 'Data Marshalling' must be employed. The concept of data
marshalling was introduced by the author in the general context
of measurement and control in 1960 (Ref.3); and was concerned
with meeting the triple problem encountered with information-
transfer in all complex systems, of identifying and extracting

critical data from a multiplicity of sources, of ensuring that
no vital data is lost, and of making '- - clearly apparent' to
the operator the '- - location and magnitude' of e.g. the
'readings' under observation.

These principles are illustrated by one embodiment of data
marshalling - the 'alarm and situation' diagram - which has as
its primary function the provision of an overall 'instantaneous'
picture of the system at any time, and particularly under alarm
conditions. Taking the general 'central processor' based type
of control system with Visual Display Unit presentation and
'hard copy' recording, the alarm and situation diagram takes
the form of a (wired) mimic board designed to give the required
'overview' picture of the system of sub-system involved.

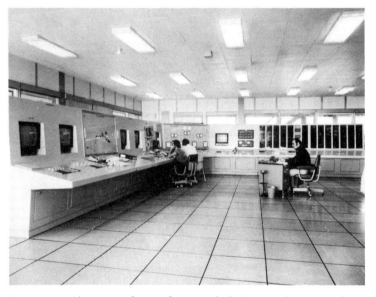

Plate 1 Master station consoles and control desks in the control room of the
Umm Shaif offshore oil field. 'Data Marshalling' A and S diagram in centre
of console in left foreground
Supplied by the British Petroleum Company Limited

Thus, for example, with a large offshire oilfield, a group
of well-heads would be shown on the A. and S. diagram with their
main pipeline runs and a pictorial representation of (say) the
block of plant at the central platform. Alarms at critical

points are shown by lamp or similar indication; the Control
Engineer thus receiving a combined 'alerting' and 'location'
signal should a fault develop. This information enables him to
select the relevant VDU pages without prolonged searching.

Plate 2 Instrument panel on a production platform in the North Sea Forties
oilfield showing the degree of complexity which may be required in a remote
site installation
Supplied by the British Petroleum Company Limited

It will be realised that it is desirable for the minimum of
detail to be shown on such a diagram i.e. visual 'clutter'
should be avoided. Furthermore the basic layout as shown is
extremely unlikely to be changed during its working life, so
that a hard-wired diagram may be used with all the advantages
given by the avoidance of equipment complexity. It may be
noted that even with the original technical proposals, made in
1960, it was envisaged that where more flexibility was needed
large-screen television might be used for the Situation diagrams;
while in Chaper 4 equivalent software-based display systems are
examined in the context of operational requirement such as 'at-
a-glance' working for both Crisis and Anticipatory Control.

It will be appreciated that the choice of the system
elements to be shown on an overview or situation diagram is part
of the overall data marshalling design; and has to be carried

out jointly between 'user' and control system designer, as
covered generally in the next two Sections in this chapter.

The other main aspect of the New Safety Regime which will
be considered briefly at this stage is the ensuring of maximum
accuracy of information 'output'. This may be stated in terms
of the information presented to the Control Engineer being highly
secure, with the instrumentation chains protected against abnormal and 'Crisis' conditions generally.

In this connection it should be noted that 'protection' in
its widest sense, has to be extended to the whole of the
installation (including control personnel) if 'catastrophic'
conditions are to be covered.

Thus one of the points raised in the UK Advisory Committee
Report referred to earlier is that of the siting and construction
of control buildings in relation to the potential dangers of
high-risk hazards such as industrial explosions. This particular issue is part of the treatment devoted in Chapter 5 to the
whole question of giving constructional and physical protection
to all sections of an installation against high-risk (hazard)
environments. Such environments can be described as 'totally
hostile'; and as carrying, at the extreme end of the scale, a
threat which is the equivalent of a natural - 'earthquake' type
- disaster.

This question of taking precautions against high-risk
incidents, such as explosion, in the design of a control room
has been used here to highlight the influence which 'environmental
engineering' must be allowed to exert on all system design, once
it has been laid down that 'crisis' level protection must be
provided.

Applying this principle more widely, leads to a re-appraisal
of the vulnerability of the various sections of the control
system to severe mechanical disturbance and, similarly, to
electrical or other disturbance of this character. This is of
significance for the instrumentation chain, and outstanding with
regard to the transducer at its input. The need for the
'complete' protection for (instrumentation) transducers has to be

8

met in two ways if highly secure information is to be obtained
under these abnormal conditions.

The first kind of protection is, as already described,
fundamentally a matter of the design and construction of the
transducer itself and of its physical environment. As will be
seen from Chaper 2, the former is far more involved than is
generally thought. This is particularly true at the 'hyper-
interface' where the parameter/electrical signal transfer takes
place; and it is at this interface that the actual measuring
process may be said to be taking place. Consequently - as
treated in some detail in Chapter 2 - it is vital that, under all
conditions, the integrity of this transfer should be maintained
(i.e. that the transducer itself should be 'telling the truth').
It becomes clear, however, that disturbances at the transducer,
especially at its hyper-interface input, will tend to lower the
accuracy of this transfer, however good the protective design of
the instrument may be.

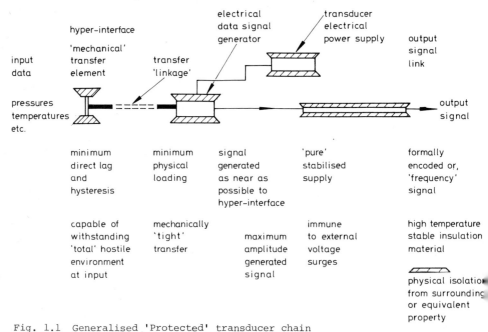

Fig. 1.1 Generalised 'Protected' transducer chain

Thus taking the extreme case of a near-catastrophic break-

down, with the transducer exposed to, say, totally abnormal
mechanical shock, it is possible for it to fail in such a way
that it continues to give a measurement indication which, though
apparently normal, is in fact completely false. Such 'false
indication' is almost certainly the greatest potential danger
which may develop with a complex system handling a large number
of measurement data sources; and the detection of an isolated
fault amongst them presents a very real problem.

Obviously 'false indication' can result from the failure of
any element in the control chain; but, as already indicated, it
is at the contact with the 'outside world', i.e. at the transducer input, where the effects of a hostile environment are most
likely to be felt. Also, the transducer is at the one point in
this chain where, almost invariably, access cannot be gained to
check accuracy under working (dynamic) conditions; while in most
cases access for state calibration by reference to a standard
is almost equally difficult.

It thus becomes evident that, under 'crisis' conditions,
the detection, and even more, the tracing, of false indication,
are key factors in maintaining overall Safety, particularly if
they can be brought into effect before a full 'Incident' develops.
It may also be noted that while the transducer is clearly the
most vulnerable point in the chain in this regard (the
'comparison-checking' which can be carried out within the rest
of the system is not directly possible here); false indication
may also take place in information display. Such false indication may be produced by equipment failure, or, on the
operational side by 'mis-reading'. Although possible -
especially under emergency conditions - as a pure human error,
the mode of presentation of information is often largely
responsible for such mis-reading.

The need to reduce mis-reading to a minimum implies the
use of data marshalling techniques, not least to achieve
freedom from spurious - unwanted - information. The display
system in itself should, of course, introduce no error in
presentation; but, as with the transducer, it is at the basic

transfer point - at the display-operator interface - that
undetected faults are most likely to occur. Again, although
failure is potentially much reduced compared with the instrumentation transducer, direct comparison-checking of the transfer
is equally difficult; and to cover the completely abnormal
conditions of a 'crisis' breakdown, it becomes necessary to
incorporate 'Independent Check' methods in the design.

The 'Independent Check' principle is one by which, for
instance, a specific measurement is covered by two completely
separate instrumentation chains, each working on an entirely
different physical basis from the other. It is applicable
to the whole of the system, and, in many respects, can be
regarded as mandatory as soon as it is accepted that design must
be for 'Ultimate Crisis' control.

Several important sub-principles are involved here. First
of all, a distinction must be drawn between Independent Check
and straightforward duplication for the circumstances attendant
upon a major 'Incident', although for both approaches, two
chains of equipment are employed. One of the main arguments for
the avoidance of duplication as such is that with two identical
sets of equipment, failure of one component due to e.g. abnormal
mechanical shock will amost certainly be accompanied by the
corresponding component in the other (parallel) set failing in
the same way. Even the suggestion of triplication can be
countered by the same reasoning, particularly if only one system
element fails completely and the other two are on the 'break-
point' i.e. are in a 'false indication' type of fault condition.
This is the classic voting problem which, whether it is being
carried out by human assessment or by some form of comparator,
essentially demands for its final solution comparison with some
form of external reference.

Bearing in mind that a totally abnormal - even hostile -
environment will exist under full Incident conditions, it is
clear that such an external reference will be suspect, either
in itself or in the sub-system associated with it. This leads
to the concept of the Independent Check as defined, where the

chances of a simultaneous, undetected, fault appearing on both
sets of equipment are greatly reduced relative to a duplicate
configuration. This can also be stated as being an arrangement
where 'common failure points' are kept to a minimum in the two
complementary systems.

Thus taking an instrumentation sub-system as an illustration,
an overall layout can be envisaged based on a primary
instrumentation chain designed to follow existing practice as
far as possible consistent with its having maximum protection
against full Incident conditions. The other chain would, in a
sense, have a secondary role. It would, of course, be designed
for maximum protection against abnormal conditions as with the
primary chain; but would have as its main design aim the
elimination of 'unknowns' along it i.e. the elimination of points
where system integrity can be lost. This loss can occur, for
example, as the result of 'break-in' effects due to noise
interference or poor 'translation-response' at a transfer point.
It will be appreciated that for any system chain, serious
departure from normal along it cannot be monitored by any
economic means; and that under emergency conditions, the
departure from normal would become so great as to nullify any
form of monitoring designed for operation under normal conditions.
For instance, excessive voltage surges can result in the whole
system becoming 'saturated' and the whole data content lost.

This inability to monitor along a chain is perhaps best
related to operation in terms of a computer-based system; and
it emerges as a definite requirement in certain cases where a
large volume of data is to be shown visually. Thus in a thesis
describing a new Computer Typesetting System, Abaza (Ref.4)
points out that '- - it would be advantageous - - to make text -
- visible to the user on a display screen (for editing and
correction) - - while it is still within the computer'. This
analogy is, of course, limited, although this particular system
is directly applicable to data presentation associated with
pictorial display as covered in Chapter 4, and is also referred
to there.

However the point of immediate interest which arises with this analogy is that when there is a delay in e.g. the printout of data in what is effectively a real-time system, then it is usually an operational advantage to gain (visual) access to the data as early as possible in the chain. Furthermore, if this access can be gained with the minimum of complexity required for translation transfer from the computer form of data to its output - in this case, visual - form integrity is more easily maintained.

This leads to the concept of information being kept in raw data form throughout the system chain; and as seen in its original usage in radar practice where the basic 'raw radar' signal chain is complemented by a second chain handling equivalent processed signals.

Taking the general problem of maintaining security of information along a control chain under full Crisis conditions, there are basically two modes of attack. The first type of approach is to adopt conventional methods and, by digital encoding, to translate the original (raw) data into a form which has inherent protection against noise and other kinds of signal disturbance. However, under the totally abnormal conditions of Crisis breakdown, high energy impulsive interference alone can reach such proportions as to obliterate completely a coded signal pattern.

Precautions can be taken against such saturation effects. These include in particular the reduction of transmission bandwidth to a minimum, and the adoption of detection methods which in themselves give protection against incoming interference. One such system is described in Chapter 3 which is designed to combine high-speed data transmission with a minimum bandwidth requirement.

Therefore, it is clear that for ultimate Crisis Control, especially when Saturation effects develop along the whole system, some form of data handling must be adopted which penetrates ('reads through') every kind of interfering disturbance.

Two methods of attack can be offered which are both based on working as nearly as possible at raw data level. Both also involve visual display; but this can be regarded as being part of the human intervention which is required if effective control is to be achieved under emergency conditions.

The first method also has the advantage of extreme simplicity where the transducer (effectively raw data) output signal is generated at high level and fed to a VDU where it appears as a sine-wave type trace. As described in Chapter 2, the value of the transducer input 'parameter' can be read by other than visual means; but in the present context, the main function of the display is to enable the operator to discriminate between the required 'value' signal and the spurious interfering signal.

The second method is essentially television-based; although in Chapter 6, 'Super-software' techniques are suggested to replace routine human observation by computer interpretation, the facility for operator visual analysis would, of course, be ratained.

The outstanding advantages of television used as a tele-metering medium or the equivalent are centred on the fact that once the intelligence has been in effect 'encoded' in pictorial form - no further degradation in its data content can take place through the system. Also, in simple terms, not only are spurious signals obvious in a TV picture, but, with the known content of the legitimate picture, it is possible to read through near-saturation interference. Furthermore complete failure of the TV system as an instrumentation chain is clearly evident; and, in comparison with other chains it is effectively 'fail-safe' in that 'no-indication' by an individual instrument within a televised group means that the instrument, rather than the remainder of the chain, has developed a fault.

The price to be paid for non-degraded raw data (transmitted instantaneously from a firmly defined original picture) is that a relatively large bandwidth is required; but this is more than justified where Catastrophic conditions must be anticipated and only limited 'Critical point' instrumentation is required.

1.2. System control - Initial planning and design

The preliminary work on any supervisory control scheme consists essentially of determining 'user' requirements in the first place; and then building up - as a joint consultative exercise between user and control system designer - the initial 'system facilities statement'. This statement is produced in tabular form under specific design area headings; and when expanded in the next phase of the work, gives the basis software requirements for computer based control systems.

One of the main results of producing the system facilities statement is that the 'user' system (plant) is defined in sufficient detail for a basic design assessment to be made of the supervisory control system. An essential part of this process is to 'separate out' the various design parameters in the control system itself and to determine how many can be dealt with in isolation and how many are 'dependent variables'. This is, of course, particularly important where the safety requirements of the main plant become over-riding i.e. mandatory for the whole system specification.

In this last connection, advantage can be taken of experience to make provision for back-up instrumentation or emergency (shut-down) control at points known to be vulnerable, especially environmentally. Similarly, to deal with the complete breakdown of a specific transducer where all output indication is lost, 'limit' alarms can sometimes be supplemented by placing local transducers in positions where they provide 'check comparison' for the initial fault condition.

Predictive analysis

Such provision can be taken still further - again in the light of experience, basically on the plant side - by looking for possible areas where consequential faults might develop, as already mentioned. In its ultimate form, this evaluation becomes 'failure mode and effect analysis'. With computer-based control systems this can either be made an 'aided' design study or

carried out at the manufacturer's works as part of the build-up
of software. Such 'works' analysis methods also involve system
simulation; and are almost certain to be continued with the
actual plant during the commissioning period on site. Furthermore advantage can be taken of the availability of the equipment
to train control engineers to deal with simulated faults which
might possibly be encountered in service.

It will be appreciated that two difficult policy decisions
have to be taken with regard to the apportionment of effort in
such analysis. Clearly, the purely design study section of the
work can take up a large amount of time, especially where a
completely new type of installation is being covered as a 'first-
time' project. Thus the first major decision is to determine
how many combinations of fault should be studied; and the second
is to establish how much work should be allocated to the design
phase and how much should be carried out on site.

Taking the first of these two points, the cost of the work
to be undertaken will soon become totally uneconomic if the
various parameters cannot be separated out as mentioned earlier,
and the 'dependent variables' are left in groups with almost
infinite design interplay left to take place between, and within,
them. This can be reduced to a great extent by using the breakdown classification of the system facilities statement.

The three basic headings in this breakdown are :
(a) Measurement and indication;
(b) Control;
(c) Monitoring and recording.

Obviously these cannot be treated as watertight compartments;
but, particularly in the light of safety requirements, it is
usually possible to isolate certain system elements, such as
fully proven 'accepted' instrumentation chains, within one of
these three sections and to treat them, in effect, as reliable
and also 'mandatory'. Otherwise the study gets completely out
of hand with virtually no limit to the number of combinations
that have to be investigated. This follows from the (implied)
assumption that every component in the total system is liable to

fail not only singly but also as part of a consequential sub-system fault.

Therefore, in practical terms, it is necessary for the control system designer to set numerical limits for the work to be undertaken in relation to his own resources, particularly number of experienced staff and scope of design aids. This is perhaps best done by estimating the scale of the 'skeleton software' requirements involved; and then referring this to the effort envisaged for the corresponding work during commissioning, and thus bringing in the second policy decision regarding the division of effort between the two areas. In any case, as already indicated, considerable reliance must be placed on experience of comparable systems; because, even when only a small number of new features are introduced into a scheme, constant reviews and 'back-tracking' become necessary. In turn this means that if any changes have to be made as the result of such interaction, the project programme slips and there is an escalation of costs which become prohibitive.

Freezing design

Also it must be remembered that excessive delays are the natural consequence of the 'domino effect' which takes place, first in the main plant system as the result of the alteration, and which in turn is 'multiplied' within the closely linked control system. These considerations apply to every aspect of system engineering design, and in the case of predictive analysis - not least with regard to software - it is more than desirable that parameters should be 'frozen' as soon as it is "safe" to do so in order to prevent completely unacceptable loss of time and a corresponding build-up of costs.

Therefore any new safety approaches which offer the equivalent of 'frozen' techniques and equipment capable of being applied to a wide variety of control systems become increasingly attractive with growth in system complexity. Not only do they provide 'cover' for the unforeseen Incident which it is virtually impossible to predict even with the most comprehensive simulation

and analysis; but also they help to give the firm design base
which is so essential for avoiding modification delay in a
project.

Although not large in comparison with the overall system
planning and design, Predictive Analysis has been examined in
some detail because of its relevance on two counts. First of all,
it illustrates how - even for a relatively small section of the
total design work - considerable design interplay must be allowed
for; and that every effort has to be made to arrive, as soon as
possible, at 'frozen' design solutions for all areas where
interaction is found. This example highlights how costs become
cumulative where preparation of software alternatives is involved;
while from the 'engineering economics' point of view alone, it
has to be taken that programming errors will be more difficult
to detect and clear with 'parallel' software packages.

The second main aspect of Predictive Analysis becomes of
major importance in its own right when 'back-up' and similar
provision is being covered in the overall design with particular
reference to Safety. Thus, if the 'protection' demanded for
Crisis Management can be provided by back-up facilities based on
experience in the design and operation of comparable installations, any section covered in this way can be frozen at an
early stage. Furthermore, taking the example of positioning
adjacent transducers to give 'check comparison', this introduces
a tried element into the system at this point on which further
design in that area can often be centred. It will, of course,
be appreciated that the opportunities to work in this way for any
given scheme are small in number; but nevertheless a search for
them is always worthwhile.

1.3 Crisis Management - User operational aspects

As noted in the later part of Section 1.1, an enormously
heightened awareness has developed of the widespread consequences
of major industrial and similar destructive accidents. This
awareness - which has appeared both nationally and internationally

- has included the recognition of the unpredictable and
inherently near, or completely, catastrophic nature of such
Incidents.

Thus, in effect, the need has been established at 'Authority
level to incorporate the methods and techniques of Crisis
Management in control systems whenever the slightest hazard
exists. The two arms of Crisis Management, viz. Crisis Control
and Anticipatory Control, have already been introduced, and are
treated in relation to existing practice in later chapters. In
this connection it should be stressed that the Crisis Management
approach and that already outlined for system planning and design
should be regarded as entirely complementary; and that the
initial work on any scheme should be carried out as summarised in
Section 1.2 with its basic task to prepare the system facilities
statement.

Once this first system facilities statement has been
completed, it is desirable to hold meetings with component
suppliers, who will have been consulted in broad terms already,
to see how their equipment fits into the scheme as a whole.
This means that the central design authority has to proceed in
three states with individual suppliers. Working from the basic
performance specification, the first stage is comparatively
simple - to establish that the various system elements are
compatible throughout the layout. This can be expressed as a
requirement for the minimum number of additions to be made at
electrical interfaces to achieve this compatibility; and to
produce the corresponding 'harmonisation' on the mechanical side.

This leads directly to the second stage, which is far less
simple than the initial process of coordination in that design
interplay must be allowed for as in the preparation of the
system facilities statement; while at this point it is necessary
to bring in the Safety requirements of the System Specification.
For this stage, they are for 'known' environmental hazards and
almost invariably demand a joint electrical and mechanical
approach. In the main, the critical points for this work are at
the interfaces already mentioned; and, generally, it is worth-

while to keep a watch on the safety implications at these interfaces during the earlier first stage meetings. This often enables preliminary work to be carried out on a standard design so that it becomes suitable for the scheme and extensive modifications are not necessary later on.

This question of modifications - which in the worst case can amount to a complete redesign - arises from the special needs of individual industries, particularly with regard to safety. However, if details of variations from existing design can be picked up early on, not only in the main plant requirements but also for the specific application; then it may well be possible to work on a standard design and produce a satisfactory variant in a relatively short time. This has obvious economic advantages; but in addition it enables samples to be submitted for safety acceptance testing in time for use at the beginning of instal- lation of the whole system. In the UK, for example, BASEEFA (British Approvals Service for Electrical Equipment in Flammable Atmospheres) is the Authority providing such a service.

Also it is easy to overlook that detail in an individual form of construction may affect quite seriously the acceptability of a unit in terms of Safety. Taking the apparently simple system element - a flameproof enclosure into which both power supply and signal cables have to be led, even the selection of a particular cable entry gland may have unforeseen repercussions. Thus if, for instance, environmental conditions are such that a definite degree of sealing is required at the cable entry, the design of the gland ceases to be purely one of mechanical support and brings in a number of factors such as relative thermal expansion and surface finish of cable sheath and gland internal bore.

Also, taking the illustration one step further, replacement of the gland entry by a plug and socket connection introduces the question of Intrinsic Safety, should provision have to be made for operation in any kind of flammable atmosphere. It is not always realised that any plug and socket assembly can constitute a potential explosive hazard if current is flowing and can be broken by removal of the plug to cause a spark. This means that some form of electro-mechanical switching interlock must be incorporated in the assembly to make the circuit 'dead' on withdrawal. This switching action can be reinforced by shrouding the plug-socket junction to give a quenching action on any spark developed during the initial separation period required before current is actually broken. This may be regarded as being equivalent to an intrinsically safe condition; but a true intrinsically safe circuit is one in which '- no spark or any thermal effect - - - is capable of causing ignition of a given explosive atmosphere' (Ref.5).

These two examples have been chosen as representative of the exploratory work that has to be done in Stages 1 and 2 to obtain system compatibility at minimum cost. They have also been chosen as being especially relevant to Stage 3 discussion where the requirements of Crisis Management are brought in. This relevance is particularly marked where completely hostile environments are encountered and various forms of mechanical protection have to be made part of overall design.

Although the transition to the much more demanding requirements of Crisis Management is almost bound to make extensive redesign necessary, it is envisaged that earlier Stage consultation would help in the economic use of existing designs.

1.4   Cost implications of crisis management

In this brief summary, the adoption of Crisis Management with its two functional divisions will be considered largely for new installations. The question of incorporating both Crisis Control and Anticipatory Control in existing installations does not arise in a direct sense because of the fundamental changes required to the systems themselves.

Thus it does appear that in most instances it would be cheaper to rebuild the whole of the control system rather than to attempt to bring in a totally new regime by carrying out modifications of the magnitude entailed in a complete change to full Crisis Control.

In an case major modifications to the main system would almost certainly be required to accommodate, e.g. new types of instrumentation.

On the other hand it might be practicable to add, for example, some of the early warning facilities of Anticipatory Control, particularly those based on simplified television methods for critical point instrumentation. Also this should be possible without causing undue dislocation during installation; and apart from the immediate gain in instrumentation capability, the calibration facility continuously afforded by TV methods is of considerable value for checking purposes during an Incident as well as for routine working.

However, taking the main issue of providing full Crisis Management for a new installation, this will be treated in terms of a computer-based system with a standby 'hot' Central Processor Unit (CPU) which is maintained in a listening mode i.e. is constantly up-dated in parallel with the active CPU.

This specific part of the overall system represents a pure duplicate arrangement for which it has to be accepted that there is no economic alternative. The major precaution which must be taken in connection with operation under Crisis conditions is to ensure that transfer from main to standby CPU should be effectively 'bump-free'. In practice this means that the changeover delay time and the performance of the system generally should be sufficiently good to maintain pre-break operating conditions. For the abnormal voltage surges which have to be anticipated under emergency conditions, the achievement of a smooth changeover demands specialised equipment on the power supply side. Usually some combination of floating and buffer stages are required to keep supply voltage changes to an acceptable minimum value; while this value has to be determined during the preparation of the Overall Specification as a system design interchange involving computer and power supply manufacturers.

It should be noted that this improved ability of the system to maintain performance in the face of extremely adverse power

supply conditions represents, in effect, a Crisis Management
facility.

In other words such an improvement has been produced by
adding to an existing system; and - quite apart from the Crisis
Management aspect - this modification has brought with it a
better system capability under working conditions.

Other examples of building upon present equipment and
techniques to give a form of Crisis Management facility includes
the adoption of a much faster sampling rate (data source scanning
rate) in a time division multiplexed (t.d.m.) system. This
greatly reduced time interval between successive scans means
that data is presented much more nearly in accordance with the
'immediate' requirements of Crisis and Anticipatory Control.
The main penalty of increasing system data rate in this way is
the corresponding for increased bandwidth on the higher speed
communication transmission channels.

Therefore, once it has been decided that the basic Crisis
Management requirement for higher data handling speeds should be
met, means should be sought for obtaining maximum transmission
speed with a given (channel) bandwidth. Methods for achieving
this maximum performance do exist; but have not been required to
be brought into use when only relatively slow speed sampling has
been accepted practice. One such 'High-speed' system, designed
specifically to have maximum immunity against heavy noise and
other adverse 'Incident' conditions, is based on a frequency
shift keying principle with 'half cycle per bit' operation; and
this is described in Chapter 3.

The remaining area in which a basically Crisis Management
facility is suitable for an existing system is that associated
with Alarm and Situation Diagram presentation. One of the most
important functions of the A. and S. diagram is to present a
comprehensive view of the system to a control engineer entering
the room which is 'uncluttered' and yet gives him a complete
picture of its operational state from which he can work. Thus
in the event of an emergency, he should not lose time, for
example, on 'interpretation' or in having to find out where the

source or sources of trouble may be located. Such a facility is, of course, capable of being made part of an existing control room system; but obviously demands a considerable number of additions and modifications if the best use is to be made of it.

The full crisis management requirement

When it has been decided that the control capability for emergencies must be 'absolute', then it becomes necessary to take every possible precaution at all points in the control chain. The adoption of this principle means that 'total' environmental conditions must be taken into account in addition to those measures which have already been outlined to take care of failure of equipment as such, whether singly or in 'knock-on' fashion.

The term 'total' environmental conditions is used to cover the combination of all 'known' hazards which can exist at a given site and the consequential effects of any form of 'Earth-quake' type of destructive accident. It is, of course, necessary to allow for such an utterly hostile environment in the original design of equipment units and the systems in which they operate; and this leads to the concept of their having to be subjected to comprehensive environmental testing, much as with corresponding aerospace components and systems.

For instance, the blast effects from an explosion are almost entirely unpredictable; and pressure distribution can vary enormously over a specific area. Nevertheless, shock-testing, for example, where the specimen is subjected to a violent deceleration force, should give some indication of how components would stand up to blast pressure transients.

However, the most far-reaching effects of an explosion are usually experienced with the large surface areas of buildings; and the need to design for this hazard has been referred to (Sec. 1.1) in connection with the siting and construction of control buildings in high-risk environments once it has been determined that 'Crisis' level protection must be provided.

With the background of this Introduction and from these last examples, it might appear that the provision of full Crisis Management demands such changes in techniques and equipment that design alone becomes prohibitively expensive, quite apart from all the development, manufacturing and installation costs of the pre-commissioning stages.

Nevertheless, it is submitted that it is possible to take advantage of existing practice in a number of fields and, by controlled selection, to produce an overall systems engineering approach which satisfies both the technical and operational requirements of Crisis Management. Some of the elements in this approach have been brought in to the preceding Section, particularly in relation to a developed version of present control system design.

Also it must be pointed out that by working from this set of 'known' bases, it should be possible to hold costs at an economic level, which cannot be expected in the same way when starting from scratch.

Furthermore, quite apart from equipment and techniques which are available in other technological areas, it will be shown in later Chapters that there are, for example, instruments which have been developed specifically for working under totally hazardous conditions, and should be suitable for adoption for service without undue modification being necessary. The background here is partly historical in that as long as the performance requirements did not have to be brought up to the standards demanded by Crisis Management, existing equipment was satisfactory; and there was no justification for incurring capital and other expenditure to provide a facility which, though desirable, was not essential.

This situation can be seen quite clearly with the preponderant t.d.m. digital systems used in supervisory control, where analogue transducers have been employed almost exclusively with the remainder of the chain being entirely digital in nature. These analogue transducers had been fully developed over the years, and with A/D conversion, were brought straight into use with the digital system as soon as these systems were introduced. Familiarisation problems with transducer installation and maintenance were thereby avoided, as were interface and other problems associated with the time-division multiplexing of digital signals from a number of independent sources.

This last example helps to show how - just as with design interplay itself - a change in one part of a system can have repercussions all along it; and that this can have corresponding repercussions on the economic side especially, as in this instance, where major capital investment is involved.

These considerations are only part of a much wider principle which can be expressed colloquially as "Never change more than one thing at a time". This apparently simple dictum, so frequently disregarded, can only be described as crucial in any form of Research and Development investigation or the equivalent; and in the context of going over to the new approach of Crisis Management, it has to be established as crucial for planning, design and implementation.

With such a blanket statement it is essential to expand and clarify it in practical terms. In the present wide-ranging connection, one of the main considerations is the 'multi-disciplinary' nature of the work, i.e. the way in which a number of sub-divisions of engineering and physics (in the fullest sense) meet at various points in the system; and have to be integrated, not least from the economic point of view.

This multi-disciplinary approach is perhaps best illustrated by the instrumentation transducer. The economic aspects of its incorporation within a scheme have already been described broadly and linked with its technical relationship with the rest of the system. However, it is at the 'hyper-interface' input to the instrumentation chain that the transducer occupies a unique position. The physical/electrical conversion which takes place at the hyper-interface can be of over-riding importance in the design of a transducer; and, as already implied, in extreme cases this can interact with the design along the whole of the system. It is not always appreciated that in 'coupling' the

mechanical input to the actual transfer mechanism in the
transducer, considerable understanding is required of the relevant physics and of instrument engineering as such, together with
electrical and electronics system design. These considerations
are taken further in Chapter 2; but it will be obvious that a
high degree of coordination is required between the various
interests, and again between them and the main system designer,
and between him and the user if a step taken in one area is not
to nullify one or more elsewhere.

Similarly, at the other end of the scale, where full Crisis
Management requirements enter into the design of buildings, the
use of 'Earthquake' to describe the hazard conditions, becomes
almost literally true. Thus apart from specialised structural
design, consideration has to be given to e.g. baffled and
filtered air conditioning, and to the demands of electrical power
supply. Also as an instance, the inherent strength of a suitably
armoured multi-pair telephone cable may make its use mandatory
for resistance to blast damage. On the other hand, its electrica
performance may be below the standard set for certain communication equipment, and it has to be established which parameters
'have to give way'.

With this brief survey as background, therefore, it is
possible to put forward certain design and other policies which
are aimed at keeping the cost of full Crisis Management at an
economic level. After all, the question of a scheme being
designed and brought into use, in effect, 'regardless of cost'
does arise when safety is made paramount; i.e. when the maximum
protection is called for in both divisions of Crisis Management
i.e. Crisis Control and Anticipatory Control.

This demand for 'Total Safety' can become a declared aim
when, for example, the go-ahead has been given for an industrial
'complex' carrying some potential risk to be sited near a
populated area or in a position where it would threaten such an
area. However, in the light of the decisions being made at
international level, and as quoted earlier, it is becoming
evident that some kind of coordinated Emergency Control system

approach may well be required for 'complexes' where there is even the remotest possibility of crisis conditions developing.

In such circumstances direct investment and associated costs will inevitably be high for a project of this nature; but it is in the indirect cost of time (also having a monetary value) that the real danger lies. Failure to produce a final Control System design where information is 'firm' and can be 'separated out' under emergency conditions, means that not only has a satisfactory technical - and particularly operational - solution not been provided; but also that the cumulative delay associated with modification and continuous re-design builds up into a prohibitively high total.

Nevertheless, it is considered that, as outlined in previous Sections, new techniques and approaches generally can be suggested which are designed to be brought in with existing basic systems to provide full Crisis Management facilities. Futhermore, with the coordination methods as described, it would appear possible to lay down comprehensive system design procedures which are economic in the use of manpower and resources despite the complexity of an individual Supervisory Control ('Telemetry') System which "- - contains a uniquely broad spread of technology" (Ref.6).

1.5 References

1. YOUNG, R.E.(1977) : 'Supervisory Remote Control Systems', Peter Peregrinus Ltd., Stevenage, England
2. Advisory Committee on Major Hazards (1979) : 'Second Report', Her Majesty's Stationery Office, London
3. YOUNG, R.E.(1960) : 'Data Marshalling', Private Communication
4. ABAZA, Y.A.(1980) : 'A Computer Display Coupler for Typographical Data', Ph.D. Thesis University of Manchester Institute of Science and Technology
5. BS 5501 (Part 7) : 1977; EN50 020; 'Electrical apparatus for potentially explosive atmospheres', Part 7 : 'Intrinsic

Safety'. British Standards Institute

6.   YOUNG, R.E.(1978) : 'Telemetry - The Present and the Future'
The Consulting Engineer, 42, No.3, London

Chapter 2

SECURITY OF INFORMATION - INSTRUMENTATION

2.1 Transducer instrumentation chains

In its function of responding to a physical stimulus and
producing an electrical data signal related to it, the transducer
is at the input to the Instrumentation Chain and the whole
Control System; and is at this position at the head of the system
where the greatest effects of environmental conditions are felt.
In addition, these conditions are such at many sites that when
the system is operating, access cannot be gained to the transducer to check, for instance, its working (dynamic) accuracy.
This corresponds to Aerospace Telemetry where the measurement/
transmission 'sender' is inaccessible during operation; and the
parallel becomes exact when a hostile environment exists at the
transducer.

It is, in fact, with the transducer more than with any other
parts of the control system, that the greatest advantage can be
taken of aerospace practice to evolve a design which is protected
more fully against abnormal and 'catastrophic' conditions. Thus
in certain applications exemplified by oil and gas wellhead
instrumentation, aerospace type transducers with high mechanical
protection against these operational conditions have been brought
into use without major modification being necessary. It may be
noted that the sealing and other techniques adopted for these
instruments were found suitable for the potential hazards of
wellhead operation, although they had been designed and developed
originally for the somewhat different environments encountered
in aerospace operation.

Design for environmental protection
Environmental protection for transducers falls into three

main areas.

First of all, there are the 'known' hazard conditions which call for example for Intrinsic Safety precautions in the presence of 'potentially explosive atmospheres' and corresponding precautions against corrosion in chemically charged atmospheres. The protective techniques used are markedly different for the two kinds of environmental threat and these are covered in more detail later in this Section.

As a sub-division of known hazards, the hyper-interface in the transducer represents a meeting-point between its input and what can be a hostile physical condition. Thus for instance in a pressure transducer, the diaphragm or its equivalent is at the hyper-interface at which transfer from parameter to electrical measurement signal takes place; and it is at this hyper-interface input point that the 'hostile' gas comes into contact with the physical interface, in this case at the surface of the diaphragm. For such applications, the diaphragm is made of some high-grade corrosion resistant alloy, and set in a stainless steel or similar body; and it should be pointed out that a combination of materials, mechanical, and instrument engineering is required for what appears superficially to be a simple design matter.

In the second division of environmental protection, random mechanical effects typified by vibration and shock have to be anticipated as far as possible in the design of the transducer. However, it is often difficult to reconcile this requirement with the basic physical construction of the instrument.

A simple example of this difficulty is afforded by the mechanical combination of wiper arm and resistance track in a potential divider (potentiometer) 'pick-off' transducer element. With the need to keep friction at the wiper contact point to a minimum, the contact is clearly vulnerable to vibration. Steps can be taken in design to counter such effects. Thus by giving a large constraint stiffness to wiper and its linkage and making it of low mass, the mechanical resonance frequency of the assembly can be brought above much of the vibration frequency range likely to be encountered. An alternative approach is to enclose the whole transducer in an oil-filled case to provide additional damping. This increase in complexity is accompanied by other problems including that of expansion of the oil with temperature (taken up e.g. in an attached bellows); while the choice of the oil itself has to be made in terms both of its mechanical and electrical properties.

Continuing with this illustration, it may prove necessary to adopt some form of anti-vibration mounting when, for instance, the assembly resonance frequency cannot be brought outside the vibration frequency range. Even then trouble may be caused by 'bottoming' on high amplitude vibration; while this effect may appear in a much more severe form under shock conditions.

This and the preceding illustrations have been given as background to the third area of environmental protection where conditions are extreme; and design is carried out in anticipation of full Crisis conditions developing from a destructive accident. For such Incident situations, where all the hazards outlined have to be allowed for, it is clear that the transducer represents a special problem in it vulnerability to environmental effects which in themselves are often difficult, if not imposs-ible, to detect. Steps to counter these effects can be taken in design; and these will be described later together with indirect means - particularly 'Independent check' methods - which provide 'parallel source' back-up information.

In this connection, it must be pointed out that although transducer enviromental engineering is concentrated mainly on physical hazards which are largely mechanical or pseudo-mechanical in nature, similar effects can result from abnormal electrical supply variations or other severe electrical fault conditions.

At this stage, therefore, it is necessary to examine the transducer instrumentation chain seen as a whole against the totally hostile environment accompanying an Incident situation. As indicated in the diagram (Fig. 2.1), even in the absence of abnormal environmental conditions this chain is vulnerable to potential sources of error such as physical imperfections in the equipment, and inherent lack of accuracy which appears whenever, for example, a finite resolution 'bracket' has to be fixed for an analogue/digital or encoding conversion.

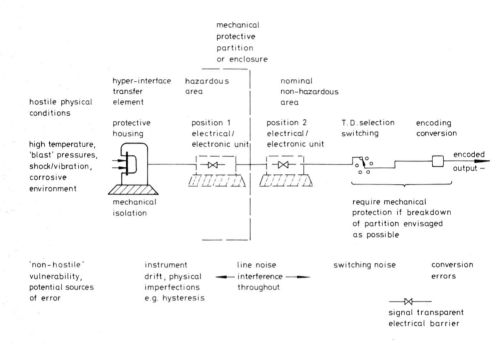

<u>Fig. 2.1</u>   Transducer instrumentation chain seen in a totally hostile environment.

Consequently, one of the first tasks of the designer is to determine as far as possible the extent to which these potential sources of error will be made more prominent in a totally hostile environment; set up at other points in the system. The latter effects will tend to be much more extreme than the former, and to lead to complete failure of one or more elements in the instrumentation chain, assuming that the transducer and its associated equipment are in close physical proximity to each other.

It should be noted that this represents one of the arguments for separating the transducer as a primary instrument from the remainder of the chain when protection is being provided against completely abnormal Incident conditions. As in all such design cases, there is a 'swings and roundabouts' aspect to the change in configuration; but, as discussed in Chapter 5, these design parameters can be broken down for a given site, and priorities established.

It is of interest that when abnormal environmental conditions are brought in to initial system planning, one of the main factors is found to be the behaviour of transistors and semi-conductors generally at elevated temperatures and the likelihood of such temperatures being developed for any significant length of time. To a great extent the magnitude of this temperature rise is determined by the type of protective enclosure adopted for the equipment and its distribution within it. It will be appreciated that even for this sub-system individual design is required for a specific set of equipment; but for the new approach demanded by ultimate Crisis Control it may be desirable to investigate the survival pattern of the transducer and this assembly if they are together and under the same environmental conditions. This is particularly important for the transducer which should be tested to 'destructive' limits if it is felt that it is possible that it will be subjected at any time to utterly hostile conditions. One aim of such testing is to provide information on the mode of failure of the component. The first piece of information required is the 'time of failure' of the sample when exposed to the simulated conditions of a total 'Hostile' environment. However, the crucial point with ultimate Crisis Control is to determine the way in which final failure is reached. It will be appreciated that from the point of view of Control, it is better for failure to be immediate (abrupt) than for a 'twilight' kind of apparent response to be obtained over a long period of decline to eventual breakdown. As already noted in the Introduction such 'false indication' can be more dangerous than no reading at all; and this applies with equal force to Anticipatory as to Crisis Control.

Nevertheless, it will be seen that although this information
on failure helps to clarify the position on unit reliability, it
is still necessary to adopt some form of parallel 'Independent
Check' system to reduce the probability of false indication to
an acceptable level under Incident conditions.

Environmental protection - General practice

As already indicated, basic protection against 'known'
environmental hazards comes under two main headings : Protection
in potentially explosive atmospheres and Protection against
corrosion in chemically charged atmospheres.

Other aspects of environmental engineering have been
touched upon earlier, in relation to the transducer and the
associated chain, notably the question of Intrinsic Safety; and
these are now brought together in a short review of the techniques and safety standards which have become accepted practice
in this field. The emphasis here is on transducers and their
associated chains; a wider treatment is given in Chapter 5 with
special reference to overall system integrity under Crisis
conditions.

In the present case of transducer the treatment is intended
to show how representative types of instrument have been designed
to withstand abnormal 'environmental' conditions, both external
to the transducer itself and also at it hyper-interface; and to
provide background to the other forms of instrumentation
described later in this Chapter.

Intrinsic Safety has already been introduced (Section 1.2)
in connection with one specialised aspect of protection against
the occurrence of sparks or other 'thermal effects' which cause
ignition of a potentially explosive atmosphere. It has been
pointed out by Weatherhead in a concise but remarkably comprehensive paper (Ref.1) that the technique of Intrinsic Safety is
designed to keep the energy contained in any spark that may occur
below the limiting value at which an explosion can be initiated.

The direct alternative to the Intrinsic Safety approach is

to allow what may be called a controlled explosion to occur where
e.g. the gas is allowed to enter an enclosure and is either
diluted by an inert gas or the flame is cooled to stop it from
propagating. It does seem that the difficulties of ensuring that
these conditions are met, particularly under Crisis conditions,
are sufficiently great - quite apart from the complication
introduced - to rule out this approach even when, as with the
transducers under discussion, the electrical energy to be
'contained' is small.

An alternative method of limiting the electrical energy is
sometimes put forward as the technique of 'Increased Safety' in
which precautions are taken in design to 'ensure' that sparks
cannot occur in the equipment. Obviously this can be a contro-
versial issue; but is is submitted that whatever the extent of
the precautions, there is always a possiblility that, in the type
of incident situation under discussion, they may not be
sufficient to prevent the occurrence of a spark. Consequently
Instrinsic Safety provision should be incorporated for these
circuits as an ultimate safeguard. In any case, as part of the
overall Intrinsic Safety approach, steps will be taken to
eliminate any chance of internal sparking (cf. the precautions
mentioned for the example given in Section 1.2).

This process is, in fact, taken to its logical conclusion,
and Intrinsic Safety is, as it were, 'designed in' to the
equipment. Thus Weatherhead shows how, knowing the various
circuit constants, it is possible to limit the voltage and
current so that the Intrinsic Safety criterion is met. There
are problems however in carrying out such a procedure when the
results of interconnection between a number of units have to be
taken into account.

The point is therefore reached when some means of isolation
has to be found for the individual 'remote' units which also
provides the necessary voltage and current limiting; and this
requirement was met by the well-known Zener Barrier Device.
Invented in the early 1960s, this device consists of a relatively
simple resistance/Zener diode network in which the main

reliability factor is the integrity of the diodes. As Weatherhead notes, a high failure rate has not been experienced with
Zener diodes in this application; and the maximum protection
would seem to be achieved with a system incorporating 'barrier'
techniqes.

Consequently it is suggested that in addition to 'Zener'
protection, the internal circuits of, say, a transducer should
be designed to add to this protection. As will be shown with a
transducer discussed at the end of this Section, further steps
can be taken to provide Intrinsic Safety features within the
instrument itself over and above the precautions taken against
the occurrence of a spark.

With the extension of Intrinsic Safety measures goes a
requirement for external protection for the equipment which is
met by the 'flame-proof' enclosure. These enclosures, produced
originally to house electro-mechanical equipment, such as
process control 'transmitters', are known as 'explosion-proof'
in the USA and Canada; but the British use of 'flame-proof'
is felt to give a more accurate indication of the form of
protection which they provide. For instance, it is usually
necessary to gain access, e.g. for maintenance, to the equipment
within the box, so that the lid-box combination cannot be made
a completely sealed assembly.

For an assessment of the 'protection capability' of these
enclosures, therefore, this question of sealing becomes a major
factor. The boxes themselves are constructionally extremely
robust, and give the mechanical protection to the equipment
itself which is so necessary under most industrial conditions.
Direct contact flame protection is also provided, certainly for
the initial period of a fire; but it is in connection with
corrosion in chemically charged atmospheres that these enclosures
are vulnerable mainly because they are not completely sealed. Put
in practical terms : 'Any of these enclosures will breathe'; and
although the ingress of corrosive vapour and gas will be slow,
nevertheless corrosion will be set up if they come into contact
with unprotected surfaces of 'susceptible' metals. Electro-

mechanical and electronic equipment in the enclosures is therefore under threat; and this applies especially to integrated
circuits where e.g. aluminium or gold are used at electrical
contact points and for interconnection.

In an excellent survey Paper, with 272 references, Schnable
et al (Ref.2) give a great deal of insight into corrosion-
failure in the thin-film metal 'structures' used in integrated
circuit and similar type components. They also give a valuable
summary of the various factors involved in the appearance of the
initial defect and the consequent development of corrosion
sufficient to cause failure of the device.

The paper classifies the corrosion effects which cause
failure of devices employing thin-film metal :
(a)   Purely chemical, where no (electrical) bias is present
(b)   Electrochemical, where bias is applied to an anode/cathode
combination
(c)   Galvanic, where dissimilar metals are in contact.

The sources of such corrosion effects are often extremely
small and 'localised', a specific mode of development being by
capillary action at a connecting lead entry point.

The paper is concerned essentially with two types of microelectronic device : 'Hermetically-sealed packages' and 'Plastic-
encapsulated structures'; and it does appear that for both there
is always a possibility that 'contaminants' - particularly
moisture - may be entrapped during manufacture or may enter
later through e.g. a pinhole defect in the device enclosure
material. It is of interest that experience has shown that
the corrosion rate in, say a hydrogen sulphide atmosphere, is
increased enormously in conditions of high temperature and
humidity. Such conditions are encountered for example in the
Middle East with temperatures of $40^{\circ}C$ and a maximum relative
humidity of 95% or greater.

Therefore although manufacturing control may ensure a very
low incidence of corrosion even in the most adverse conditions,
there is always a sufficient element of doubt to suggest that,
for maximum protection, additional precautions should be taken

with electronic equipment of this general nature.

One such approach is to put some form of sealant coating on the board assemblies; but it would appear that, even with the various application techniques that have been introduced, the possibilities of pinhole or similar defect developing still remain.

In this connection it is of interest that similar structural defects are experienced with the inorganic 'passivation' layers usually of a 'glassy' character, which are employed with integrated circuits. Organic coatings, including epoxy and silicone materials, are also used with integrated circuit type assemblies; but again mechanical defects can develop, exemplified by cracks in the coatings which occur where there is no metal immediately under that part of the passivation film.

The remaining form of protection is indirect in principle, and is used with the equipment enclosures as described earlier. The technique adopted is that of 'sacrificial absorption' where e.g. fine turnings - commonly the sacrificial metal is copper - take up the corrosive agent in an active layer on their surface. In practice the turnings or similar high surface area ratio material are in an open dish in the bottom of the enclosure cabinet, and are replaced at routine intervals, typically of the order of 6 months.

Finally, some features of existing types of transducer will be described which provide environmental protection at the instrument itself sufficient for defined hostile environments.

First, there is the physical problem of corrosion which has been treated above, both in the general electromechanical context and for microelectronic forms of construction; and also mentioned with regard to the use of stainless steel and other corrosion-resistant materials. In this connection, some warning has already been given of the difficulties that can be encountered with the multi-disciplinary interactions in transducer design; and a number of these considerations immediately arise with the choice of materials for a transducer diaphragm as an apparently simple example.

These considerations include the reconciliation of corrosion resistance requirements with the other metallurgical properties of the various materials used in the diaphragm region, particularly with regard to differential expansion with rise in temperature. This in itself becomes an important design issue when elevated temperatures in a Crisis condition are being taken into account. With extreme temperatures, buckling of the diaphragm could take place with ultimate mechanical failure leading to the opening-up of, for instance, the plant pressure line to the atmosphere in the event of a major pressure surge. Furthermore with any of the forms of strain gauge/diaphragm combination, there would be electrical failure, partial at first, and yielding a dangerously misleading signal, followed by complete material deformation but not necessarily accompanied by a total loss of signal, i.e. a 'fail-safe' condition would not be reached.

Such catastrophic failure is anticipated by the designer in establishing his initial 'worst-case' design parameters which can be checked from the known characteristics of the materials involved. Thus, assuming these characteristics are for the extreme ranges of temperature and other parameters, a 'structural' solution can be reached which can be used as a basis for the next stage of design for total Crisis conditions.

The main factor to be brought in at this stage is the dynamic response to rapid changes in applied pressure. It can be stated that cases have been experienced with diaphragm-type transducers where the dynamic response to sharp pressure transients was entirely different from that obtained by static testing; and that this effect appeared to be a function of the material used for the diaphragm.

Such 'long-hysteresis' effects are difficult to detect in service. In any case the abnormal pressure changes necessary to produce them are unlikely to develop except under fault conditions. However, it is under severe fault conditions that this measurement information is most required; and it is in the light of this requirement that significant work is justified on joint dynamic and metallurgical testing.

It must be added that a number of techniques have been developed to clear incipient sources of trouble during manufacture. These techniques include prolonged pressure cycling and 'accelerated aging' of materials which themselves have been subjected to full quality control.

Another generic type of transducer - the linear voltage differential transformer (LVDT) - lends itself to working in hostile environments, and the protection methods which have been adopted for some of the members of the LVDT family are of special interest in their own right and as compared with the techniques already described.

In one advanced transducer of this type, made in the UK, the ceramic-insulated titanium coil former is contained within an 'environmental' case which is hermetically sealed by electron beam welding. The sensing coil itself is wound with anodised aluminium wire so that high temperature-rated insulation is provided as the result of the anodising action to the exclusion of physically susceptible forms of insulation. As special dry lubricant is used for the rubbing surfaces set up between the armature core and the inside of the coil former tunnel. The operating temperature range is given as $-55^{\circ}C$ to $+450^{\circ}C$.

A differential-inductance (four-winding) bridge variant of this transducer was designed for operation in an extremely hostile environment of intense neutron irradiation at temperatures up to $500^{\circ}C$. The anodised aluminium wire of the winding was carried on a ceramic former and sprayed with a ceramic coating. Boron-free materials were used throughout.

In a somewhat different sphere, that of flow measurement, a variable reluctance transducer of American origin (Ref.3) becomes of special interest for Crisis Control. This is not only because of its relative mechanical simplicity, but also because it can accept corrosive and contaminated fluids at high pressure (maximum stated as 5000 p.s.i).

The flow valve is derived directly as a function of the displacement of a 'spool' armature held in the flow stream, and

not indirectly as, for example from pressure-head measurements.
The spool is held by two opposed helical springs in its zero-
flow null position relative to the two transducer coils, themselves mounted coaxially on the flowmeter body. Square-root
extraction to give a linear law flow measurement is achieved by
contouring the bore of the metering section so that the
required flow-area variation along the common axis is obtained.

Vibrating cylinder transducers

Another form of transducer which has certain advantages for
working in hostile environments is the vibrating cylinder type
(Ref.4) despite its dependence for operation on the critical
mechanical dimensions required for the basic sensing cylinder
and on its comparatively complex 'energising' sub-system.

In these primary transducers, working on the mechanical
resonance principle, the sensing cylinder is maintained in
constant oscillation as a 'mass-spring' system with the resonant
frequency f given by $\omega = 2\pi f = \sqrt{k/m}$ where k is the stiffness and m the mass. Thus, variation by the input parameter of
the effective (total) vibrating mass or of the spring stiffness
will result in a corresponding change in frequency. It thus
becomes possible to carry out measurements of density and
pressure; and it should be noted that density measurements can
be made on gases as well as liquids (including slurries - by
virtue of the straight through flow path).

The ability to measure density of a moving liquid is of
particular importance in the 'independent-check' context in that
this facility enables mass flow readings to be obtained from
one instrument; the measurement characteristic being given in
terms of change of frequency from a specific variation in density.
This represents a completely different physical basis of
measurement from the more usual 'two-transducer' method of
deriving mass flow; and with its low number of system elements,
the vibrating cylinder transducer offers considerable advantages as the alternative instrumentation chain.

Also among these advantages (as with the balanced

oscillator type to be described in the next Section) is that
conversion to the varible frequency output takes place at the
hyper-interface itself with no other elements interposed in the
chain. It should be pointed out that in addition to this
transfer being 'immediate', the resultant electrical signal is
being generated at high level. This is particularly desirable
in the presence of abnormal 'break-in' noise and interference
generally, both from electrical and mechanical sources. Also
with regard to vibration-induced noise, because of the high
mechanical Q of the resonating element (representative value
around 3000 at 1 kHz), there is some immunity to external
vibration at the much lower frequencies usually encountered in
practice. Nevertheless the implied warning given earlier in
this chapter with regard to the difficulties of protection which
must be expected with intense vibration must be repeated here in
view of problems such as 'mode-jumping' in the vibrating element.

However, it is suggested that on balance the necessary
precautions can be taken to deal with totally abnormal vibration
and associated conditions, the vibrating cylinder transducer
could be used in the 'Independent check' role, notably for mass
flow measurements.

Balanced oscillator (dual chain) transducer principle

It can be stated that the development and the basic design
of this transducer principle and the range of instruments based
on it were carried out from the beginning to meet the requirements of full Crisis Control. This applied particularly to
environmental protection, and with it, the maintenance of
minimum error under abnormal and 'catastrophic' conditions.

These Balanced Oscillator (dual chain) transducers (Ref.5)
have a 'parameter-dependent' frequency output derived as a
difference (beat) frequency between two matched oscillators
themselves 'tuned' by high resistance moving probes. These high
resistance (e.g.Michrome) probes move differentially within
their sensing coils and form the input to a dual chain electrical
system based on a balanced mixer. Consequently. with this

43

symmetrical system, compensation for wide changes in ambient
temperature and in supply voltage takes place by virtue of the
inherent balance of the system.

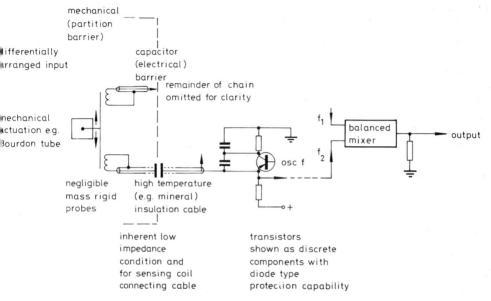

Fig. 2.2 Balanced oscillator transducer system showing 'designed-in'
intrinsic Safety features

Full advantage is taken of the properties of the Clapp
oscillators (Ref.6) which are used as a matched pair at the
hyper-interface input to the system. With the series resonance
operation of the Clapp oscillator, the consequent low impedence
condition, and its 'high-C' feature, wide variations can be
accepted in the design of the sensing tank coil and of the probes.
In turn, coupling between coil and probe does not have to be
tight so that electrical loading is minimal, which adds to the
frequency stability of the transducer seen as a mechanical-
electrical combination.

Mechanically the probe is not in contact with the sensing
coil and does not have to be driven over, or supported by, a
bearing surface, nor does a lubrication problem arise. Also the
probe itself can be made of negligible mass, so that the effects
of vibration and shock are extremely small; while the corrosion

effects encountered with variable reluctance, and especially with variable resistance, transducers are virtually eliminated.

However, the greatest attraction of the sensing-coil/ oscillator combination for extreme environments is that the former can be installed remotely from the 'electronic' unit with a low impedance (coaxial cable) connection between the two. The importance of this possibility of isolation, especially in terms of Intrinisic Safety, will be apparent. With suitable 'barrier' protection, both electrical and mechanical the coil can be situated in the hazardous area and be made effectively neutral with regard to energy being available at this point to initiate a spark. It is also possible to arrange some Zener diode type barrier protection within the electronic chain following the transducer element, especially if discrete components are used.

An equal contribution can also be made, on the data handling side, both for detecting 'false indication' and for providing the transducer output signal which can be read out in the most adverse circumstances.

In the range of transducers based on this principle, the measurement frequency output is obtained from known types of instrument 'actuation' systems. For instance, an equivalent probe 'pick-off' may be fitted to a Bourdon tube actuator to replace a sensing potentiometer (potential divider). Consequently by retaining the conventional pointer and scale, visual system check calibration of the transducer becomes possible with pointer 'read-out' and corresponding transducer measurement signal being produced by the same Bourdon tube actuator. Thus by using one of the Closed Circuit Television methods of Section 2.3 remote system check calibration of the transducer can be carried out with no loss of accuracy in transmission of the visual indication.

Advantage can be taken of this facility to make system check calibration of the transducer as a routine matter whenever desired, and without having to gain access to it. This last aspect is obviously of particular importance for a site where known hazards exist or where the installation of the transducer renders it completely inaccessible.

It must be pointed out that the term 'system check calibration, is used to distinguish it from 'absolute' calibration.
In the latter case, this is a process carried out in comparison
with an independent reference standard; whereas the main object
of 'system check calibration' is to determine quantitatively if
there has been a deterioration in the accuracy of transfer
through the transducer system from hyper-interface to output
'reading'. However, with a stable form of actuator, as exemplified by the well-tried Bourdon tube, a convenient method of
remote observation is provided which, for day-to-day operation,
is a close equivalent to an 'absolute' calibration facility. The
main objective here, of course, is to detect 'false indication',
and particularly to obtain early warning of the commencement of
failure.

It should also be pointed out that this is not a pure
Independent Check system as defined with two physically separate
instrumentation chains. In this case there is a common component;
the measurement input actuator; and this particular arrangement
is chosen to give the comparison check of the transfer chain as
described above.

As far as the measurement output signal is concerned, the
balanced oscillator transducer shares the advantages of the
vibrating cylinder instruments in being generated at a high level
'at' the hyper-interface.

However, the main advantage of this type of output signal is
seen under 'catastrophic' conditions where, say, electrical
interference and noise disturbance generally, tend to obscure
the legitimate signals especially as far as computer-based data
processing is concerned, and where random bursts of impulsive
noise are coming in at saturation level.

In such circumstances, the beat frequency or equivalent
output signal can be read "through" the noise by visual means,
and using the human intelligence to discriminate between wanted
and unwanted signals. Such a procedure would, of course, only be
adopted in an Incident emergency, and for 'Critical point'

measurements (see Section 2.4). In practice, the measurement of
frequency (and hence the corresponding known value of the
parameter e.g. pressure) could be made on a 'short-gate' counter,
the reading being taken when interference was absent as indicated
either on a CRT or even by a listening check. In extreme cases
of 'blanket' interference, a reading can still be taken directly
from the CRT trace (which can be built up against the random
noise by trace integration, as with radar); or, adopting an
early technique of the UK Met. Office, by taking the reading of
a calibrated beat frequency oscillator arranged to produce
Lissajou figures with the incoming frequency on the CRT.

Plate 3 Long term stability tests on Balanced Oscillator Protected
transducer system, CRT trace showing raw data frequency output.
Supplied by Manchester Polytechnic

2.2 'Independent check' Transducers

The Independent check approach has already been introduced
as a general principle and largely in terms of system design and
operation.

At this point, however, two specific types of transducer wil
be described which each work on an entirely different physical
basis from the instruments they would complement in an Independen
check installation. Also it would appear possible to adapt them

for operation in hostile environments without any change in
fundamental design being necessary; and, in fact, the first type
to be considered - the 'single-ended' variable reluctance probe
- was first developed for telemetering from inside internal
combustion engines (Ref.7).

Variable reluctance probe transducer

Designed originally to measure clearances between a station-
ary (bearing) surface and one moving relative to it, this probe
has two immediate applications in the present context :
(a) To detect and measure very small relative displacements
(resolution is claimed to be of the order of 10μ in.) Such a
measurement facility could be used with large pressure vessels to
obtain derived values of temperature and pressure change from
differential movement (strain) produced by abnormal build-up of
stress within the vessel. Such Critical point instrumentation
depends for its effectiveness on a multi-disciplinary analysis at
the measurement information 'take-off' point (Chapter 5); and it
is in this connection that the engineering methods associated
with the second - 'Washer' - transducer to be described are of
direct interest.
(b) In the second application, advantage can be taken of the
small deflection characteristic of the probe transducer to use
it with diaphragm 'actuation'; and with this configuration, to
evolve a design which has a high over-pressure capability
provided for abnormally large surges by constraint against a
heavy disc.

The mechanical arrangement of the probe is centred on an 'E
and I' magnetic circuit of cylindrical form which encloses the
single pick-up coil. The E solid of revolution of this magnetic
circuit is made up of a number of open loops of soft iron wire
which are machined off to give a mating surface to the corresponding area against which relative movement takes place and
which acts as the I. The ability to withstand high temperatures
is determinded largely by the soft iron wire core material. This
has a Curie point of $770^{\circ}C$ which permits operation up to $600^{\circ}C$

on this score. With this basis, and by using vacuum filling and
baking techniques with a special glaze, a form of ceramic
construction is obtained which enables the whole transducer unit
to be used at this elevated temperature.

Pressure sensitive 'washer' transducer

Designed for the measurement of petrol engine "fluctuating
cylinder pressures", and of Japanese origin, this piezoelectric
ring transducer is inserted between the spark plug and its
cylinder block seating. Its maximum operating temperature is
given at $150^{o}C$ and its resonant frequency to 16 kHz (Ref.8).

The method of clamping, involving tightening to a specified
torque value, would appear to be applicable to making e.g. a
machine foundation bolt head a measurement take-off point.
Similarly the ring transducer could be placed under a bolt head
in the flange of an enclosure under pressure. In the first
instance, early warning of the development of machine vibration
should be obtainable; while in the second, similar warning of
violent (detonation type) pressure changes should be given. Such
occurrences would be associated with Incident conditions; but it
is under such conditions that such 'instrumentation for emergency'
might well provide information which would not be forthcoming
from more 'conventional' types of transducer.

The choice of measurement information take-off point would -
as with the variable reluctance probe - demand appreciable
investigation time before settling the final transducer
installation details.

2.3 'Independent check' CCTV systems

The use of closed circuit television for comparison checking
has been mentioned in Section 2.2 in connection with two
instrumentation chains which have a common transducer actuation
input.

In the examples which follow, the two instrumentation chains
are 'looking' at the same physical parameter; but are entirely

independent of each other in that the measuring devices at the
inputs to the two chains are completely different from one
another. This true independent check working is being achieved,
where one chain is transducer-based but the other handles its
measurement information as a video signal derived from a 'meter
board' or equivalent digital measurement display. The outstanding feature of the TV data handling chain used as a telemetering
medium, is that once the intelligence has been in effect 'encoded'
- in this case presented in suitable pictorial form - no further
degradation in its accuracy can take place through the system.
There is, of course, some loss of accuracy associated with the
transfer between the physical parameter and the digital
(numerical) read-out on the meter dial; but with a well-tried
type of, say, pressure gauge, this can be regarded as negligible,
certainly for the Independent checking role. The errors which do
arise are relatively small and can be explained by instrument
drift and physical imperfections such as hysteresis. It is,
however, important to give as much protection as possible to
these meters for operation under Catastrophic conditions, so that
there is maximum probability that their readings will be maintained in these circumstances. This leads to the 'instrument
box' concept where the meters and the TV camera are housed in a
robust flameproof type enclosure. This enclosure, with the
'flying-spot' system to be described, is made light-proof to aid
in keeping power requirements to a minimum.

An inherent strength of the TV based instrument chain is
the way in which, in effect, it removes a disturbing, and often
dangerous, uncertainty in operation when a transducer instrumentation channel is 'lost' and it is not clear whether the trouble
is at the instrument or on the transmission link. With TV
working, the reading of the meter as a 'visual transducer' will
be transmitted to the control-point, and any purely instrumentation fault will be evident. Should the transmission link
itself fail, this also is evident with the complete loss of
picture. The value of this information is clear when the TV
system is being used for Independent check purposes and a fault

develops on the complementary channel under catastrophic conditions.

It will be seen that such a TV scheme can be made comparatively simple using a conventional camera with black and white working. However, particularly when only an isolated meter is to be observed, a low-cost solution can be offered by 'flying-spot' method. Analysed in terms of installations based on this principle, certain advantages can be claimed both with regard to cost and to safety as compared with a conventional TV camera system.

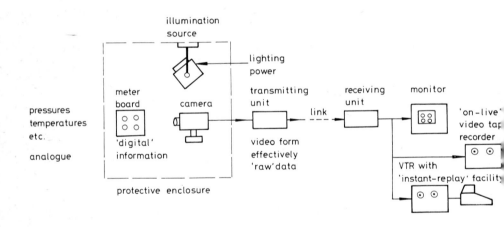

Fig. 2.3A CCTV telemetering - conventional camera

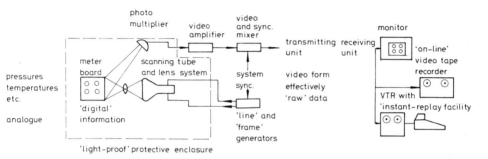

Fig 2.3B CCTV telemetering - 'flying-spot' reflective scanning

Thus with an early airborne research installation (Ref.9) a small dial instrument panel was scanned through a wide aperture lens by a non-interlaced 200 line 50 field per second CRT raster. The scanning optical system was completed by a small multiplier photocell. One of the reasons making for lost cost with the CRT/photocell combination is that no light source is required to illuminate the subject being televised. This has two attendant advantages for use under hazardous conditions. One is that no power has to be brought into the area to feed lamps; and the second concerns the vulnerability of such lamps to vibration and shock - this mechanical problem does not arise.

With regard to mechanical vulnerability, all the airborne equipment for this scheme was subjected to environmental testing as part of a programme of gaining data on potential reliability; and it was found that the component most under doubt - the scanning CRT - was surprisingly robust when under test for vibration and for shock.

2.4 'Critical point' instrumentation

The idea of providing special instrumentation facilities at certain Critical - key - points where, for example, high mechanical stress can develop in a structure, has already been introduced in connection with Crisis Control.

It will be realised that the importance of gaining accurate measurement data at these Critical points justifies the use of Independent check methods where applicable, and of employing highly secure instrumentation methods where these can be devised.

One instrumentation principle which offers such security is given by a combination of transducer and television techniques. In such an arrangement, the approach suggested for the Balanced Oscillator transducer (Section 2.1) for common actuator working is taken to another level so that physical phenomena, such as deformation under stress, are 'picked-up' mechanically in such a way that they can be viewed by TV methods. The instrumentation system which results has a fundamentally simple mechanical acuator which introduces virtually no transfer error to the numerical indicator scale; and where this 'digital' reading is transmitted directly to the observation point.

The heart of a system of this kind is the actuator; and for the measurement of pressure-induced strain, a near-ideal solution is given by the long established extensometer (Ref. 10).

An alternative method of measuring such small displacements is afforded by a micrometer clock gauge (Ref. 11). In this particular case, the clock gauge plunger movement was produced by an inclined ramp.

Finally extensions of this CCTV based approach to take advantage of colour television and other techniques are given in Chapter 6.

2.5 References

1.   WEATHERHEAD, D. (1977) : 'Intrinsic Safety', Measurement and Control, 10, London

2.   SCHNABLE, G.L., COMIZZOLI, R.B., KERN, W., and WHITE, L.K.
(1979) : 'A Survey of Corrosion Failure Mechanisms in Micrelec-
tronic Devices', RCA Review, 40, Princeton, NJ

3.   INSTRUMENT DIVISION OF LYNCH CORPORATION : 'VR-30 indicating
flowmeter', CC 7601.1, Detroit, USA

4.   POTTER, P.N., (1972) : 'Vibrating element transducers'.
Colloquium on novel types of transducer, IEE, London

5.   YOUNG, R.E. (1975) : 'A new universal digital transducer',
75 REMSCON ('Remote supervisory and control') Conference, Newport
Pagnell

6.   CLAPP, J.K. (1948) : 'An inductance-capacitance oscillator
of unusual frequency stability', Proc.IRE, pp.356-359, and
pp.1261-1262 (ROBERTS, W.A.)

7.   YOUNG, R.E. (1965) : 'Piston engine telemetry', Ind.Electron.
(March)

8.   COMPUTER ENGINEERING LTD. (1980) : YG-11 'Pressure transducer', Hitchin, England, 1980

9.   YOUNG, R.E. (1955) : 'Flying Television', Hawker Siddeley
Review, London

10.   GOODMAN, J. (1926) : 'Mechanics applied to Engineering',
Ninth Edition, Longmans, Green and Co. Ltd., London

11.   YOUNG, R.E. (1961) : 'Analogue telemetry equipment and
systems - Pt. 1', Electron. Eng. 33, No.395, London

Chapter 3

SECURITY OF INFORMATION - DATA TRANSMISSION

3.1 The requirement for 'Instantaneous' Information

Particularly for 'Critical point' instrumentation, there is a requirement for information to be presented to the Control Engineer with the minimum delay i.e. to be as nearly as possible 'Instantaneous' in nature.

This requirement arises in two ways :
(a) For Early Warning, especially with regard to the development of Crisis conditions
(b) For effective handling of an actual Incident.

At first sight this would appear to imply that two types of data transmission (communication) system would be required to handle the routine control data and the 'Instantaneous' crisis management data respectively.

However, even for a relatively small plant, the provision of two separate communication systems is uneconomic, both with regard to cost and to complexity. A still more important reason for avoiding two systems is that by planning an 'overall' single scheme, it is possible to provide for one part of the system to act as a 'back-up' and a form of monitor for other sections.

This does not mean that duplication of communication channels is not required; but that, if at all possible, there should be a common policy behind design, aimed above all at giving the maximum degree of protection to all parts of the system and at minimum cost.

There are obviously a number of design decisions involved in implementing such a policy; but it does seem possible to meet the over-riding requirement for protection, and yet provide adequate communication performance. Techniques have been developed which enable e.g. high-speed data transmission to be achieved over

limited bandwidth circuits; but these have not been called into
general use in the past in the absence of the firm demand which
can be anticipated if Crisis Control measures are made paramount.

Communication performance and overall design

In the circumstances of an 'Earthquake' type disaster, taken
as the ultimate design criterion, mechanical protection of
equipment takes very high priority. At the same time, the
transmitted information must be given maximum security with
regard to freedom from interference and from other sources of
error such as demodulation trouble at the receiving end of the
communication link due to excessive attenuation along it.

At the beginning of the planning and design of a communication system, there is, in theory, a choice available between
line and radio links. However, even when extreme hazard
conditions do not have to be taken into account, there are a
number of 'constraints' which militate against the use of radio
links (Ref.1). These include propagation difficulties for line-
of-sight working in any form of built-up area, and power supply
problems at remote sites and where repeater stations have to be
employed. In any case, however, there is often no alternative
to line working because of the difficulty of obtaining a frequency
allocation for a radio link.

There are occasions, nevertheless, when a radio link has to
be used often on an operating frequency which is far from ideal.
Such installations are covered in Section 3.4.

However, it will be taken that the system design parameters
in the present treatment are based on a line (cable) link; it
being possible to state from comparative studies that this
represents a satisfactory approach, not least with regard to
environmental protection. Thus, with the multi-pair balanced
cable which is the optimum on the electrical side, the mechanical
strength can be made extremely high by known forms of construction.

Taking first the question of physical construction, a
typical multi-pair cable as employed for the special video link

described in Section 3.3 would have its polythene-insulated conductor pairs held symmetrically within further (spacing) insulation surrounded by steel tape armouring acting as an external screen. By substituting continuous heavy steel wire armouring for the tape, high composite strength is obtained; and at least one occasion can be quoted where such a cable withstood the blast from an explosion in close proximity to it without any damage to the conductors being experienced.

On the electrical side, the balanced pair configuration offers high inherent protection against interference break-through and noise generally. It is, however, necessary to maintain this electrical balance in the equipment terminating individual circuit pairs for maximum noise rejection to be achieved.

It must be stressed that this type of cable, is essentially of 'telephone-grade' low bandwidth rating. That, in addition to telephone working, it can be used to provide circuits for high-speed digital data transmission and for the video link already mentioned, is made possible by the special techniques described in the next two Sections.

Allocation of communication resources

Within the framework that has been outlined, there are basically two divisions of communcation channel. The first of these - for Critical point instrumentation - is essentially a 'continuous' channel permanently allocated to one source of information.

The second division is associated with time division multiplexed control systems, while, with computer-based digital working, are accepted as being standard for many applications of remote control. In this division, data source addresses are scanned 'in turn' on an 'interrogation-reply' basis, where these sources are grouped in blocks of addresses according to priority. The low priority sources are interrogated in accordance with a predetermined scan cycle, so that the updating time for an individual source is equal to the scan cycle period. With the

selected high priority sources, interrogation is much more
frequent, with their block of addresses interlaced within the
overall system scan period. The updating time for these high
priority sources is therefore much reduced, and, in terms of the
information coming from them, they fall between the 'continuous'
- Critical point - channel and the low-priority 'slow-scan'
channels.

Bearing in mind that the multi-pair cable is common to all
these channels, the possibility suggests itself of being able to
switch say, the cable pair taking low-priority information, to a
higher priority function during an emergency when it was known
that extra information was required urgently from that area.
This principle could be taken further by switching such a cable
pair to a 'continuous' channel if more critical point information
could be made available on it for 'Crisis' operation.

There would, of course, be technical complexity involved in
the switching process in that terminal equipment on the cable
would have to be changed when going over from, e.g., a low
priority transducer signal to the output from a television based
source. The decision as to whether such a changeover facility
should be provided would be largely an operational one; but
might well be justified in making it possible for additional
Critical point instrumentation to be installed after experience
in service had established the need for it.

The final concept of making such changes during actual
operation when attention had to be concentrated on a Crisis
area is discussed in Chapter 6 in terms of 'Super-software'.

3.2 'High-speed data transmission on line links'

For rapid transmission of digitally encoded data over a
minimum bandwidth link, it is necessary to send the highest
possible number of bits per unit bandwidth.

This demand has two main components :
(a) For the maximum speed of transfer of new information for
system up-dating as discussed, and particularly for high priority

58

data sources.

(b) For the minimum bandwidth to be utilised to transmit information. This is an 'economic' requirement; but in the present
context, the most important advantage of keeping 'occupied bandwidth' to a minimum is the consequent reduction in noise
interference.

The sytem developments to be described are designed to meet
the performance criterion of 'minimum cycles per bit' for f.s.k.
(frequency shift keying) type working.

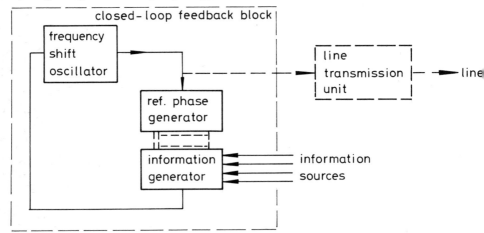

Fig. 3.1 Low bandwidth f.s.k. data transmitter, basic loop

In the first period of development, the structure of the
modulated waveform was made such that the marks and spaces
corresponding to the applied pulse modulation contained an equal
and integral number of their respective signal (keying) frequencies. In the generation of this type of waveform, two main
requirements have to be met to ensure that the modulated signal
is held to the 'equal and integral' distribution of cycles in
the waveform. The first of these requirements is that transitions between marks and spaces should be at zero axis crossover
points; and the second that there should be a minimum harmonic
content at the two signal frequencies.

In practice, this waveform is generated in a closed-loop
feedback system containing the frequency shift oscillator. The

oscillator control loop is closed through an 'information generator/reference phase generator' combination which is fed from the output of the frequency shift oscillator. This output, i.e. the modulated signal, is sent through a line transmission unit. The receiving loop is of the same general configuration, but includes a gating subsystem to provide stepping-on signals for a circuit block consisting of the reference phase generator and address register.

Further work has been done to reduce the number of 'cycles/ bit' for these f.s.k. systems, mainly at the University of Manchester Institute of Science and Technology (UMIST); and it can be stated that the final objective of 'Half-cycle' working has been reached with, as a representative example, transmission rates approaching 6000 bauds having been realised with nominal 3 kHz circuits.

With the increase in transmission speed achieved by the change from one cycle to half-cycle per bit comes a major change in system approach. The former 'index' is, by definition, the limit for a symmetrical waveform (about the zero axis); so that this characteristic of the waveform is lost immediately this limit is passed. Nevertheless, with the asymmetrical waveform, the two main features of the earlier waveform are still retained but with some modification made necessary by the loss of symmetry.

Thus transitions between marks and spaces are still arranged to be at zero axis crossover points, this axis being a key reference both for modulation and demodulation action. The transition between half cycles as seen in the modulated waveform is made as smooth as possible. This aids in the demodulation process and also helps to reduce harmonic content; and in the same way, discontinuities and abrupt changes in slope in the waveform are kept to a minimum. This result is achieved by the use of what are sometimes known as 'coherent' modulation tech- niques and by waveform 'shaping' methods including the use of digital filters.

3.3 Video link on low bandwidth cable

As indicated earlier, it would be possible to transmit
television pictures for, say Critical point instrumentation, over
the telephone-grade cable suggested as being suitable for all
communication purposes in a control system capable of dealing
with Crisis conditions.

The early installation (Refs 2 and 3) to be described was
designed to give acceptable picture quality over an 8.35 km
(5.2 mile) route length of ten-pair cable without intermediate
repeaters. By adopting the original British 405 line picture
standards, relatively low cost camera and monitor equipment was
adopted for black and white working.

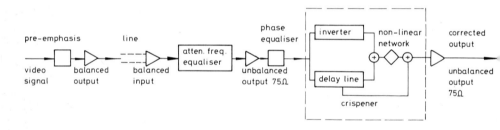

Fig. 3.2 Low bandwidth cable video link, schematic

In this case, the 0.9 mm diameter conductors were polythene
insulated, and with steel tape armouring giving an overall cable
diameter of some 22 mm. With a conductor resistance of 26.4 Ω /km
(44.2 Ω / mile) at 15°C attenuation at audio frequencies was
0.744 dB/km (1.20 dB/mile) and crosstalk between pairs better
than -80 dB measured on site. Attenuation reached a value of
some 80 dB down at 1.2 MHz with an unequalised frequency response
approximating to the form 1/$\sqrt{}$ (frequency).

After equalisation, a uniform response within 2 dB was
obtained up to 1 MHz. The overall response was about 6 dB down

at 1.2 MHz with relatively sharp cut-off thereafter.

A further improvement in high frequency response was obtained by adding transmission pre-emphasis amounting to 10 dB with a 3 dB point at 200 kHz. The necessary phase equalisation was done at the receiving end, some $0.5\mu s$ correction being given at 1 MHz.

Finally, a 'crispener' was incorporated to sharpen up fast edges in the video waveform by speeding up their rise times in a nonlinear network. An input filter in the crispener unit was used to extract the fast edges from the equalised video waveform for feeding to the nonlinear system. This filter operates by signal subtraction referred to a wideband delay line to give a Gaussian form response. After passing through the nonlinear network, the artificially sharpened edges were recombined with the original (incoming) video waveform, the delay in the filter being compensated by introducing a corresponding delay in the main video path.

3.4   Radio links

The question of the constraints which act against the use of radio links in hazardous conditions has already been touched on. There are, as noted, the primary constaints of frequency allocation and aerial siting in built-up areas or the equivalent, and the restrictions associated with power supply.

Nevertheless, if it becomes clear that there is no alternative to a radio link, the following points are relevant when dealing with the provision for Crisis control.

In certain cases it is possible to get some relaxation on frequency allocation if transmitter power is kept low and the radiated beam is extremely narrow. Both of these requirements imply the use of microwave frequencies - one of the main factors operating here is that the cost of an aerial support structure capable of producing the desired narrow beam becomes prohibitive with longer wavelengths. Even with the smaller microwave aerials, however, the aerial structure has to be made highly rigid if the beam is not to be deflected from its line-of-sight arrival at the receiving aerial (and vice versa) either by local earth tremors or by gusts of wind.

It should also be remembered that microwave links suffer propagation attenuation, dependent on wavelength (increasing with frequency markedly below 10 cm wavelengths), heavy rain being a major contributor to this.

It can be said that the use of a narrow, directed, beam helps to reduce incoming (radio) noise, which nevertheless would tend to be high under catastrophic conditions.

Some light is thrown on these effects in relation to power line carrier (p.l.c.) working in an authoritative Paper by Sealy and Senn (Ref.4). Thus it is stated that external interference (on the p.l.c.communication circuit) can arise from lightning, line corona discharge and also from circuit breaker and isolator switching surges. Also some interference is experienced on the p.l.c. circuits from the transmitted power itself; and it is of interest that this varies between a.c. and d.c. links. Thus, for the same weather conditions and voltage stress, interference from d.c. lines is less than from a.c. lines. Furthermore, whereas the interference from d.c. lines is reduced by the presence of rain or wet snow, the reverse is true of a.c. lines. Dry snow increases the d.c. interference level. With regard to fog, d.c. interference is comparatively unchanged while a.c. interference can be appreciably increased. These results are given for the p.l.c. transmission frequency range of 40-600 kHz (l.f. and m.f. bands); but they are of value for close proximity 'blanket' considerations at microwave frequencies.

Finally, where a radio link must be used but demands for power have to be kept to an absolute minimum, the possibilities offered by 'quiescent' and 'low drain' working may represent a practicable solution for routine t.d.m. operation.

In quiescent working, only the receiving section and the alarm logic in the outstation are 'alive' in the absence of scan interrogation from the master station, with minimum demand placed on the power supply. Full demand is developed when the out-

63

station is activated either to signal an alarm condition to the master or to accept interrogation.

For low drain operation, the outstation is 'on' continuously taking perhaps 5 W mean power from, for example, batteries charged by solar cells. This can be achieved with low power consumption logic, and usually involves switching on the transmitter as in quiescent working proper.

In the general context of keeping radiated power to a minimum, the question of 'radio hazards' at the transmitter in a flammable atmosphere does arise.

Considerable experimental work has been done in this area; some of the evidence is conflicting, and further work is undoubtedly still required, but it does appear that for a spark to be developed at a dangerous point, the equivalent of a transmission line stub or open loop, must exist at that point, which is of a physical 'length' which is resonant at the radiated frequency.

3.5 References

1.   YOUNG, R.E.(1977) : 'Supervisory Remote Control Systems', Peter Peregrinus Ltd., Stevenage, England
2.   LAYZELL, M.C.(1969) : 'Closed circuit television link via telemetry cable', Water and Water Engineering, London
3.   YOUNG, R.E.(1970) : 'Industrial telemetry', Wireless World, London
4.   SEALEY, T., and SENN, W.(1972) : 'Telecommunications associated with e.h.v. transmission systems'. Conference on Telecommunications for the Service Industries, UMIST, Manchester

Chapter 4

DATA PRESENTATION

4.1   The basic presentation requirments and data marshalling

For all forms of engineering control it is necessary to
present the operational information to the control engineer in
such a way that :
(a)   This information can be assimilated almost sub-consciously
(b)   Delay and error (particularly ambiguity) are kept to a
minimum.
These requirements become increasingly important as the
speed with which action is demanded becomes greater i.e. the more
nearly full Crisis Control is approached. Thus for an emergency,
the control engineer should have been fully in touch with the
operational position at all times through Anticipatory Control
facilities; and when he had to exercise over-riding control, was
able to do this without a mental 'search-time' being necessary.
To enable these requirements to be met poses a major problem
for the designer of the data presentation equipment, because,
quite apart from the design of the visual displays themselves,
Data Marshalling methods have to be devised to deal with the
'streaming' of the information to be displayed. 'Streaming' may
be defined as a data handling process in which the information
is first selected and then directed to the most appropriate point
of presentation. It is clear that the maximum amount of
streaming decision-taking should be built into the system during
its original design. This does, however, conflict with the need
for operational flexibility which becomes particularly important
when Crisis Management conditions are entered.
This need for flexibility can be seen as arising from one of
the main functions of data marshalling in producing a particularly stream : to extract only 'useful' information and yet not

miss anything which might bear upon the control action in progress. It will be realised that during a major Incident, significant changes will take place in the 'rating' of information, that is data which was regarded as no more than of a routine 'housekeeping' nature may suddenly become a key to the way in which the emergency has developed.

The term 'housekeeping' has been borrowed from the Aerospace world to bring out the need to detect as soon as possible when, say, such a routine temperature measurement has become an indicator of the beginning of the development of a dangerous plant failure. Advanced methods of picking out the onset of these threatening conditions as part of the application of 'Super-software' are put forward in Chapter 6; but these are intended for complex control systems where it has been established that so much Critical point and Independent check instrumentation generally is required that human analysis and assessment may have to be supplemented by computer-based methods.

However, apart from these exceptions, it is vital that the directness and physical simplicity of Independent Check instrumentation should be retained in its presentation of the information which it has derived. This should be on an individual ('tied') basis using television or equivalent techniques to 'relay' and display this instrumentation information.

By their nature, these displays - less diagrammatic in character than the conventional types of semi-pictorial VDU or other multi-source display - require some interpretation of the information as presented - and therefore there is a further potential difficulty in coordinating them with VDU 'page' working, for example.

On the other hand, it is possible to visualise at least one method of operation where the functions made possible by the Independent check principle could be performed as a self-contained process complete in itself, and then 'indexed' on the the VDU pages. These functions - carried out largely in terms of 'raw data' - are essentially to provide unique Critical point instrumentation, to test that specific instrumentation chains are 'telling the truth' and to watch for, and even to anticipate, their failure, and finally to provide reserve and back-up information facilities; and it would seem quite feasible to correlate the results obtained with VDU and similar data presentation as used in many control installations.

It must be stressed that the design of such a data presentation system must always have as its background the total scheme which it will serve. Thus, for instance, at a relatively early stage firm 'plant requirements' for data presentation have to be established so that the individual design and layout of display consoles and control desks can be begun.

In this case of Crisis Management and as shown in Chapter 5, it is even more important that constant reference should be made to the basic plant requirements for data presentation design, when what are, in effect, two Control Regimes are being brought together in the final scheme.

The first of these Regimes may be described as applying to 'delayed-in-time' control operation and is associated with 'standard' computer-based systems which may be regarded as a form of closed-loop control system in which the loop is closed through a human operator. Working from telemetered plant measurement and indication, the control engineer takes action only after he has absorbed the incoming information, i.e. there is a delay in response in that he is not in a position to react immediately. This is the case even when a fault condition develops because the information is not received immediately and in general, requires appreciable time to check in terms of display and presentation as a whole. It is only when the second Regime - control operation in 'real' time - is entered, that it is possible for response to be much more rapid; and the form of overall data presentation employed is a major factor in obtaining this result.

It will be clear that the introduction of a real time element in basic control sytem design will have implications on the operational side; and it is in both these connections that it is valuable to draw a parallel with comparable Air Traffic

Control system (Ref.1).

4.2 The air traffic control analogy

Both the plant (real time) Crisis Control scheme and the air traffic control system have the same basic structure in relation to the human operator. Each is engaged in data handling and can deal internally with large volumes of information; and with CRT display, selected data can be presented with virtually no delay to the operator who is interposed in an over-riding (intervention) position in the overall control loop.

Storage of this data is required on a relatively large scale for both of these systems. With plant control, this is exemplified by the requirement to store and coordinate their transfer to display and recording. These facilities are found in the basic control system. In the ATC case, similar storage is seen for speed and other flight path information held for individual aircraft for the duration of their flight path through the controlled air space.

Taking the ATC direct loop, radar (data) signals are displayed to the operator in near-pictorial fashion and in real time; this takes the form of conventional radar-map presentation with aircraft height information available. Interpretation of this display information is therefore required before instructions can be sent back to the aircraft under control by radio-telephone. Again delay is involved in this interpretation process; and it is acknowledged that the design of the display can make a major contribution to reducing this delay, and, as noted earlier, to keeping error and ambiguity in reading to a minimum.

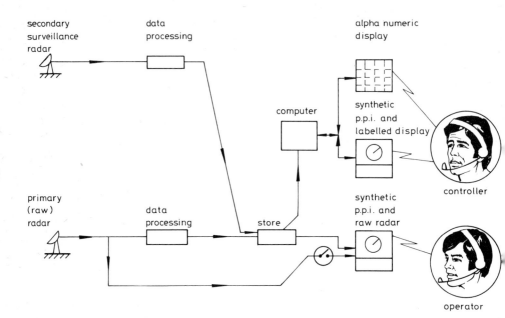

secondary
surveillance
radar

data
processing

alpha numeric
display

computer

synthetic
p.p.i. and
labelled display

controller

primary
(raw)
radar

data
processing

store

synthetic
p.p.i. and
raw radar

operator

Fig. 4.1 Air traffic control - primary/secondary radar system - information flow diagram

With air traffic control centred on radar information, it is of interest that the plan-position indicator (PPI) has been retained as the heart of the data presentation system (Ref.2); and that with computer processing, PPI type displays are used both at the operator's and at the controller's positions. From the simplified diagram it will be seen that the former shows a combination of raw (primary radar) data and derived, synthetic PPI signals; while in the latter the synthetic PPI information is complemented by computer 'labelling', with an auxiliary 'alpha numeric' display providing alternative data presentation from the same information source.

Although the parallel is not exact, it would appear that there are various aspects of the overall system which are closely related to the combined plant control scheme with its two systems for routine and crisis management respectively. One of the main points of difference is that the main display at both operator's and controller's position is basically the same ('synthetic PPI') whereas with the proposed plant control scheme, the usual presentation method for routine operation is of a different character from that for full crisis management. This difference is also seen in the fact that operation on both ATC 'channels' is essentially in real time; but with plant control there is, in effect, a 'spread' of time scales ranging from that of the routine Control regime to the second regime with Crisis operation under emergency conditions.

However, it is in this last area that a valuable precedent exists with the ATC system in that it has proved possible to bring together and correlate operationally two sources of data through individual display for each information channel. As already indicated, such correlation is the ultimate aim for plant control with full Crisis management in operation; and it is suggested that relevant experience has been obtained with air traffic control which could be fed into the design of certain sections of plant control data presentation. This appears to be especially applicable to operation under the rapidly changing conditions covered in the next Section.

It is, however, with regard to the more general system engineering principles of 'Independent check' methods that most interest lies in the ATC system. With the arrangement shown, the equivalent of cross-connections make it possible not only to 'get inside' the two information chains at selected points along them, but to do this with raw radar signals as a reference.

4.3 The aircraft (piloting) control analogy

This particular analogy has been chosen to give a brief indication of how a pilot is helped to deal with rapidly changing operational demands corresponding with those encountered during 'Incident' conditions.

In the first place, if at all possible, means have to be found to 'separate out' the various pieces of information coming in to him in real time (in data marshalling parlance, to stream them); and to make sure that this is done without saturating him with too much data.

The potential dangers arising from having too much information presented at any one time were recognised in the Aerospace
world many years ago. For example, Fogel (Ref.3), in a paper
published in 1959, pointed out that, with the general tendency to
build up pilot's instrumentation aids to an unprecedented degree
of complexity, too many 'reading' actions were demanded of him in
too short a time.

This leads to the simple but nevertheless fundamental
concepts of making the presentation device to be read of such a
form that the reading time becomes minimal; and of arranging the
instruments in such a configuration that they 'fall into place'
naturally for the observer to read them.

To achieve these aims demands a psycho-physiological approach
which in turn requires close attention to the form e.g. of the
displays in themselves, particularly with regard to their not
being distracting and holding the observer's attention too long.
On the other hand they should permit 'at-a-glance' reading, a
concept which raises a number of issues in its own right,
particularly to question whether such a facility is really
required.

In the pilot's case the answer lies in the periods when he
is under maximum operational stress i.e. when the aircraft is
taking-off and, even more markedly, when landing. These conditions of stress correspond to those envisaged for an Incident
situation where the control engineers are dealing with a fully
developed 'Crisis management' emergency; and this is one of the
main reasons why the present analogy has been pursued as being
entirely relevant.

Reverting to the second high-stress period - the landing
phase - it is suggested that the parallel with Incident control
is closest when this phase is being carried out in zero visability
under ground control; and extremely 'rapidly moving' control
information has to be exchanged between ground and air, while,
in addition to being engaged in this 'interactive' operation, the
pilot has to 'fly' the aircraft which does not involve reference

to his instruments, and to which task he can give only a portion
of his time.

Thus, for conditions such as these, 'at-a-glance' reading
does seem to be inevitable, and with it comes the clear impli-
cation that displays should be in analogue form. This, in practical terms, means that the 'clock' type of meter should be used
in these high stress circumstances for the display of critical
information.

Such an assertion may seem surprising at this stage in
development of digital displays; but it can be stated that at the
time of writing (1981) 'clock' flight instruments were still the
preferred form of presentation effectively the world over. In
confirmation, a caption to a photograph showing a complete
preponderance of clock instruments contained the following:
'Flight instruments (and radio installation) on a typical modern
general aviation executive aircraft ---' (Ref.4).

From the point of view of Crisis control, it can be pointed
out that an important advantage of such analogue displays is that
- provided they are maintaining a continuous reading - they show
'trends' in the parameter being measured, something which can be
vital in the 'early warning' detection of e.g. a dangerous
pressure build-up. Also, in response to the suggestion that
reading accuracy is lost because of its 'non-digital' nature, it
can be stated that a conventional clock dial with seconds hand
can be read in two widely separated places at the same instant,
and the divergence between the two readings will be at worst, 2
seconds, and usually zero. This is, of course, with trained
observers, but it is assumed that this would apply to the
personnel involved in this kind of control activity.

The emphasis given to these analogue displays is intended to
demonstrate how 'at-a-glance' principles have been met with this
particular form of presentation; and to indicate how it might be
possible to extend them to other display techniques especially
to the 'Alarm and Situation Diagram'. It does not mean that this
policy should be adopted for the main plant displays, where for
routine operation, digital techniques must remain in use on

economic grounds, quite apart from the purely technical reasons
for their adoption.

4.4 The crisis management man/machine interface

In both divisions of Crisis management - Anticipatory
Control and Crisis Control - the display of the information
derived e.g. from Critical point transducers has to be in analogue form for the reasons already given. With 'frequency' type
transducers, this output must be shown in its original - 'raw' -
form as a CRT waveform as described for measurements to be made
under full emergency conditions; and this facility should always
be available. However under somewhat less adverse conditions,
this transducer information can be shown on dial instruments,
derived in this instance as a.d.c. analogue signal.

In cases where the local dial reading of the transducer are
relayed as TV signals to the display point, the final 'picture'
is of the same kind as the derived meter reading; and it becomes
apparent that these compatible indications can be brought together at, say, a central display console. However, even with
the limitation necessary for the number of Critical point
instrumentation channels, there is obviously a requirement to
identify them so that 'search time' is not involved for Crisis
operation. This implied demand for a means of identification
suggests that the philosophy associated with Alarm and Situation
Diagrams might be adopted in this case.

With such an Identification Diagram there are basically two
possible approaches for setting up the 'labelling' required.
The first - applicable to relatively small schemes only - would
be a hard-wired arrangement with perhaps some degree of
allocation/selection carried out by manual switching. The second
approach would be essentially software-based; and would either
demand its own microprocessor or would entail dependence on the
main central processor computer. In either case this provision
could not be regarded as a true emergency measure.

However with sufficient accumulated experience of operating

a given plant, it might well be possible to establish a 'two-
level' system of Anticipatory Control where full Critical point
instrumentation techniques would be applied at the first level;
but a software-based early warning system would constitute the
second level. As described in Chapter 6, much more developed
'Independent check' and similar facilities would enable this
second level system to be integrated with the main control system;
and this basic arrangement could be followed for an interim stage
where data on the performance of the early warning system itself
could be acquired.

4.5 'Incident' control recording

Following on the concept of building up data on the performance of the computer-based second level system, as outlined above,
is the possibility of recording the operation of the whole Crisis
management group of systems during an emergency.

One of the main disadvantages of 'pictorial' data as
obtained e.g. with TV type relaying is that conventional-oriented
data handling and recording cannot be used. Thus although video
tape recording can be used for the pictures, data reduction
methods, as generally understood, cannot be used to extract the
data in a form suitable for handling in a computer.

Nevertheless it is suggested that 'Instant replay' techniques might be employed with standard video recording to check
back on key parameters as they took place before the actual
onset of Incident conditions; and it seems quite possible that
this procedure might be developed with experience into a useful
operational tool. In practice two VTR machines could be used
for this work : one 'on-line' machine together with the second
machine for the Instant replay function which at any given time
would hold a record of the last 30 minutes of operation to be
called up for investigation of events at any specified point
in that period.

Plate 4A Video Tape Recording 'Instant replay' control position
Supplied by Ampex Corporation.

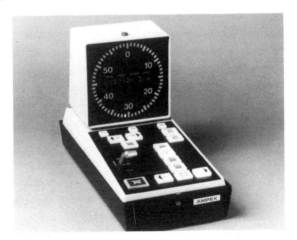

Plate 4b Close-up of 'Instant replay' slow-motion controller unit
Supplied by Ampex Corporation.

Finally it would appear desirable to have available a
printed events log for such a control operation which could also
be used for identification reference for the video recording.
The recording system for producing such a log would be designed

to be completely flexible, and to enable editing and correction to be carried out on the 'text while it was "still within the computer".

These requirements are met by a 'Computer Display Coupler' (Ref.5) developed originally to work interactively with a general purpose - mainframe - computer to give a high resolution type-setting system operated by a standard format keyboard.

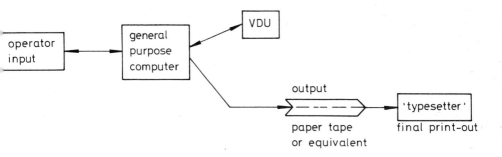

Fig. 4.2 'Computer display coupler' - basic typesetting application

Apart from the ability to work with a mainframe type of computer, this system is of direct interest in the present context because it is designed to provide fast generation of high definition 'pictures' and - especially important - fast selective erasure for editing modification.

The Display Coupler itself is based on a 16 kilobyte RAM storage capacity and a microprogrammed display processor. The 'picture' (the text) is generated as a 'display file', and when editing is complete, the file contents are transmitted to the main computer or an equivalent for final print-out processing.

4.6   References

1.   IEE (1967) : Air Traffic Control Systems Engineering and Design', IEE Conference Publication No.28. London
2.   EASTWOOD, E.(1979) : 'Radar engineering : progress and prospect', Electronics and Power, 18, London

3. FOGEL, L.J.(1959) : 'New Instrumentation Concepts for Manned Flight', Proc.Inst. Radio Engrs. 47, 1978

4. HEARNE, P.A.(1981) : 'Avionics - The Third Dimension', Aerospace, 8, 3, London

5. ABAZA, Y.A.(1980) : 'A Computer Display Coupler for Typographical Data', Ph.D. Thesis, University of Manchester Institute of Science and Technology (1980)

Chapter 5

OVERALL SYSTEM DESIGN FOR CRISIS MANAGEMENT

5.1 Assessment of environmental conditions

The general design of control systems for operation in a
known hazardous environment and under emergency 'Incident'
conditions has already been covered, directly in Chapter 1, and
in relevant Chapter sections in other parts of the book.

In the present Chapter the main aim is to concentrate more
on design for utterly hostile environments and for operation
under 'Earthquake' conditions where, in an extreme case, only
the Crisis control system has not been put out of action.

With these terms of reference, the first step in initial
design planning is to determine where there is a permanent
Hazard environment at outstations in a Supervisory Remote Control
Scheme. An obvious example is the potentially flammable
atmosphere at an oil wellhead. In such a case data on the
conditions at the wellhead will have to be provided by the
operating authority; and with this information, design can begin
in consultation with the designers of the main plant and the
'User'.

With a control room design, the position is not so straightforward, particularly if it is to be situated in the middle of
a large industrial complex where the buildings are devoted to
inherently dangerous process work.

Two stages of assessment are involved here. In the first,
the procedure is as for outstations in determining the known
hazards in that area, and in carrying out design for safety in
consultation with the main scheme authorities on what may be
called a local basis.

In the second stage, however, an assessment has to be made
of the effects resulting, say, from a destructive explosion in

any part of the main site area. Clearly, to investigate every possible combination of primary explosion and secondary effects would be completely prohibitive in expenditure of time alone; but it is suggested that the following evaluation programme, though simple, gives enough information to enable planning decisions to be taken e.g. the degree of physical protection to be given to the interior of the control room building.

To 'reduce the variables', it should be possible in the first place to establish the maximum number of alternatives which should be considered for the actual site of the control room. In practice the proposed layout of the plant and the nature of the site itself taken together, should reduce this choice to, say, three at the most. This assumes that this evaluation is being carried out during the first design studies of the main plant, and that a final choice has not been made.

With these three tentative sites as a basis, the next stop is to ascertain the location of all maximum risk areas; and then to look at the potential danger they 'present' at each of the three control room sites. It should be pointed out that the assembly of this information is not primarily associated with the siting of the control room; but is done to provide 'worst case' environmental design data which it is more than desirable to have available to feed into the design of the control room as early as possible. This information will also have reper-cussions on the construction policy adopted for the main control room building; this aspect is considered in Section 5.4.

At the same time, as part of the coordination of outstation and control room design, a similar hazard study should be undertaken for the outstation sites, but with regard to the 'isolation' of the equipment by a hostile environment. Put as inability to gain access to the equipment during operation under these conditions - the Aerospace 'remoteness' situation - this isolation makes it necessary to introduce the various techniques such as check calibration and variants of critical point instrumentation, as already discussed.

Thus the results of the hazard study, once available, can

be put together with those obtained from direct instrumentation
studies to introduce a final instrumentation scheme for that
specific section of the installation; and can be combined with
the data resulting from the Critical point evaluation of
Section 5.2.

5.2 Critical point and similar instrumentation requirements

For the study demanded in this instance, reliance has to be
placed on the designer of the main plant to indicate where
critical point instrumentation is needed and the relative priority
of individual requirements. It will be realised that it is
necessary to keep the number of these instrumentation points
below a 'system capacity limit' for that particular scheme.
Because of the nature of the observation methods which have to
be adopted for the individual readings necessary with these
transducers or the equivalent, a limit has to be placed on their
numbers to avoid saturation.

Also, further reliance has to be placed on the main plant
designer to determine where the true critical point is on, say,
a pressure vessel. Thus taking this example, although the
maximum stress concentration may be taken as being at a sharp
radius cross-section; much larger values may be found elsewhere
due, say, to the transferred tension of a holding-down bolt.
This is, of course, an extreme case; but such considerations are
responsible for the way in which, for instance, multiple strain-
gauge instrumentation is placed on test structures - sometimes
on the actual structure itself - to determine the point of
maximum stress.

Consequently, if it is decided that it is utterly vital
that the true critical (pressure) point should be determined in
this way, a further decision may be taken to extend these tests
beyond the conditions resulting from internal applied pressure
to dynamic testing to represent serious external mechanical
disturbances. in most cases of this kind, this would have to
be done with a scale model to enable standard capacity environ-

mental testing equipment to be used.

Such equipment and techniques are described in the next
Section, and it should be added that for outstations in a hostile
environment, it would appear desirable to carry out such testing
on their electronic equipment; and, for that matter, on the
corresponding master station equipment if there is any likelihood
of this equipment being subjected to e.g. violent shock.

5.3 Environmental testing

In this Section, advantage is taken of aerospace experience
and methods to give something of the background to the present
position on environmental testing; but, as has been stressed for
other areas where aerospace practice has been quoted, the utmost
caution is required when drawing parallels. Thus in comparing
these two technological fields, it becomes evident that their
respective environments are widely different and considerable
'engineering discretion' must be exercised in moving across from
one field to the other.

Nevertheless successful transfers of testing techniques can
be made; and it is of interest that the recent announcement of
the 'updating' of an existing UK environmental testing facility
(Ref.1), built primarily for aerospace applications, included
references to non-aerospace work which could be undertaken.

As indicated in Section 5.2, two main divisions of environmental testing would appear to be required in the present context.
The first would be concerned with the structural behaviour under
e.g. violent shock; and the other division with electronic
equipment, and particularly transducer systems, and would cover
a wide range of environmental conditions, notably temperature
and pressure variations.

Another type of environmental testing - for 'at-sea'
conditions - was developed in the mid-1960s in the USA (Ref.2)
which has relevance for offshore oil and gas installations.
Tests were carried out with 1/10th scale models in a specially
equipped tank; and with simulated currents of up to 10 kt and

winds of 'typhoon' strength up to 150 kt scale equivalent. Gusts could be simulated by pointing the axial fan blowers. simulating gale force conditions, away from the model and then suddenly redirecting them at it. Waves could also be produced in the tank either in sinusoidal or 'unstable' form.

In the wellhead instance, results of this kind could be of value when installing special instrumentation at the water line on the rig; and for providing data for the investigation of wave-induced vibration on the platform support structure.

It is however, the facility for producing winds of gale force and above which is of most direct interest in the present survey. First of all, with suitable 'funnelling' of the simulated gusts, it might be possible to gain some information on the effect of blast pressure waves on buildings. In the second application of these techniques, data would be sought on the mechanical deflection of aerial support structures caused by gusts. This is a specialist requirement associated with the maintenance of a line-of-sight path between transmitting and receiving aerials as covered in Section 3.4.

In both of these cases, with models being used, it is necessary to take into account the 'aerodynamic' effects associated with scale - e.g. Reynolds number - which affect accuracy. As far as general practice is concerned, it is usually found that wind tunnels, by their very nature, do not cover 'gust' conditions; so that such investigations must be regarded as requiring special provision for them to be carried out.

In contrast, 'Shaker' vibration testing, using electro-magnetic (moving coil) drive units, have been a prominent environmental testing tool for many years. Thus a water-cooled unit having a peak force rated at 28 000 lb was available in 1968. This particular unit was quoted as giving a maximum displacement of 2.5 cm (1 in) peak to peak over a fequency range of 5-2 kHz with an amplifier having an output of 140 kW. In 1981, a corresponding unit working over an extended frequency range of 'a few hertz up to 5 kHz' was quoted as being able to produce up to 18 000 lb of peak thrust.

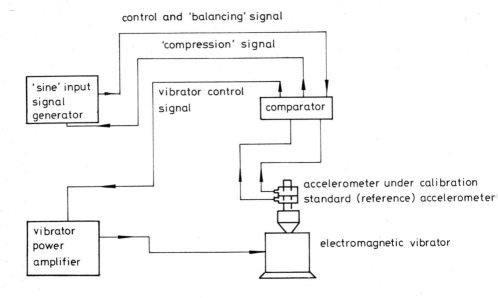

<u>*Fig. 5.1*</u> *Electromagnetic 'Shaker' vibration testing system as used for*
accelerometer calibration

The vibration control servo-system is usually arranged to
give a choice of constant acceleration, velocity, or displacement
with varying frequency; and with suitable programming, it should
be possible to obtain some information on effects of shock on
equipment which is too big for the conventioal 'drop-testing'
system. In this last type of testing, the equipment under
investigation, mounted in an instrumented carrier structure,
develops acceleration under free-fall conditions, and then is made
to decelerate violently by some form of 'anvil' buffer. A degree
of control of the deceleration period is obtained by choice of
materials at the anvil; an alternative is to use hydraulic
shock absorber techniques.

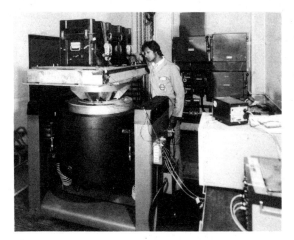

Plate 5   *Electronic package under vibration test on large electromagnetic drive unit*
Supplied by British Aerospace Dynamics Group, Stevenage

Therefore, for investigation of response to shock, drop-testing can be employed for relatively small pieces of electronic and similar equipment; but for larger assemblies or structures, high-capacity vibration testing has to be adopted. Data on shock in this case, however, is not obtained so directly as with drop-testing.

The remaining division of environmental testing interest here is concerned more with long-term investigation of the resistance of materials and equipment to extreme and cyclic variation of climatic conditions. Such testing covers the effects of dust, sand, organic mould and sea water, in addition to the more usual environmental conditions of temperature, air pressure and humidity. Climatic chambers for this work are made of considerable internal volume, a typical example (Ref.3) having a diameter of 5.5 m (18 ft) and a length of 6.7 m (22 ft).

Plate 6   *The interior of a modern climatic chamber, its size may be gauged by the Land Rover undergoing icing tests*
Supplied by British Aerospace Dynamics Group, Stevenage

5.4 Siting and construction of control buildings

As stated in the Introduction (Section 1.1) :"- - -' protection', in its widest sense, has to be extended to the whole of the installation (including control personnel) if 'catastrophic' conditions are to be covered". It was also pointed out how, in taking precautions against high-risk incidents, such as an explosion - specifically in the design of a control room -" - - -'environmental engineering' must be allowed to exert - - - a major influence - - - on all system design, once it has been laid down that 'Crisis' level protection must be provided".

With this as a background, two main issues arise. The first is the question of the form that the control building should take; and the second is the essential decision as to whether the Control Centre for the scheme should be kept as a single control room or split away in physically separate master and sub-control rooms.

Taking the first of these issues, certain assumptions can be made, the most fundamental of these being that, in the present context, blast of a high order is the major hazard which has to

be countered. It should be added that a destructive industrial
explosion will almost certainly set up earth tremors, but the
indications are that attenuation through the earth combined with
the 'solid' form of construction to be described below is
sufficient to bring this particular risk below that of blast. It
must be stressed that, for obvious reasons, this information is
empirical and does not include any data on the results of
naturally occurring earth tremors.

Numerical values for the 'overpressures' which may be expected are given in the Second Report of the UK Advisory Committee
on Major Hazards (Ref.4) and its references, already quoted in
Chapter 1; but, as suggested earlier in the book and particularly
in the preceding part of this chapter, environmental engineering
tests with scale models would seem to be necessary to gain a full
appreciation of the conditions to be expected for a specific
site.

Turning to the 'hard' control building which is implied
here, two types of construction can be put forward. One is the
reinforced concrete 'block', which, of sufficient strength, may
have an area weight distribution which is excessive for
foundations with certain kinds of substrata. The alternative is
a form of control tower built into a 'surround' building of
reinforced store rooms and, possibly, inner offices.

One partial alternative is to replace the surround building
by 'blast walls' designed in this case to 'stand alone' i.e. to
have no constructional backing behind them. An indeterminate
factor is the 'reduced pressure' (near-suction) wave which it
is thought may possibly follow the main pressure shock wave.
This largely hypothetical effect is mentioned, however, because
it would seem remotely possible that some tests might be carried
out in the climatic chambers described earlier, where, to
simulate the high vacuum conditions of outer space, pumping can
bring the pressure down to 5×10^{-6} torr or lower. The time to
reach such pressures from atmosphere is usually of the order
of 24 hr, so some transfer mechanism would be required to give
the abrupt surge required.

With regard to such control buildings, air-conditioning may
be taken as being mandatory with windowless construction,
involving some kind of baffle system in the ducts to reduce the
effect of blast. It is suggested that it might be possible to
produce a combined baffle and flame arrester assembly for these
ducts, which would give the necessary blast pressure reduction
and also give protection against the propagation of a dangerous
flame (Ref.5). The main objections to flame arresters are that
they offer some resistance to the air flow and that they require
maintenance cleaning at regular intervals. Nevertheless, where
control rooms have to be sited near a potential explosive hazard,
it may well be that the only effective form of blast pressure
baffle is the flame arrester; and quite apart from the flame
propagation aspect, reduction of sudden pressure build-up
justifies investigation. It should be added that with an
explosive shock wave of this kind, 'building dust' is blown
through the air conditioning ducts by the blast, and does take
time to 'clear'. There is no need to stress the importance of
having adequate standby power available to supply the air
conditioning equipment and control room lighting in these
circumstances.

Multiple control room installations

The decision as to whether there should be one or more
Control Rooms would appear to depend to a great extent on the
degree of complexity of the control system itself. The argu-
ment for having more than one control room is that if the first
is 'knocked out', operations can be transferred to the second;
but unless they are separated from each other by a sufficient
distance on the site, the explosion that put the first out of
action will almost certainly do the same to the other.
Consequently, with this physical separation between them,
changeover is made extremely difficult even for a small plant
under emergency conditions; and for this, with a large installation, it becomes virtually impossible with the large number
of interconnections that have to be set up between the two
control rooms.

Nevertheless, some of the bigger schemes are so designed
that certain local ('actuator') control loops are set up in the
event of failure at the Master Station, and system control of
them is taken away from it. It might thus be possible to
transfer them back to external control by the second Master
Station, arranged especially for this.

In general, however, with a large centrally controlled
scheme, the combination of control and main (plant) systems must
be treated as a whole, and sections cannot be shut down indepen-
dently from the main flow of the plant.

The other exception is, of course, when an installation
consists in effect, of a number of independent plants (i.e. is
not a 'flow' layout) and segregation of control rooms on an
individual basis becomes permissible.

5.5 Wide coverage 'development' recording

In bringing a new type of process control or other
centralised control system into full operation, data must
be gathered in a variety of areas on service performance.
This applies especially during the pre-commissioning stages,
and particularly to soft-ware 'debugging'.

Therefore when a scheme is being brought into use which
embodies full Crisis management facilities, it would appear that
two entirely separate data recording programmes would be
required. However, it is possible to envisage a stage-by-stage
procedure which would enable these two programmes to be kept
together, and which would also have the advantage of producing
coordinated methods ready for final operation.

Working on the basis of a 'standard' computer-based t.d.m.
'address/reply' plant control system, certain improved Crisis
management facilities can be added to it in its own right which
can be put into use for recording this management data. This
assumes that the main plant software is being finally 'proved
out' and modified on site.

First of all, with TDM serial working, it is taken that the

system (sampling) scan is maintained continuously. It is not always realised that because of the responder action associated with the scanned channels, a maintained succession of replies shows a channel to be serviceable. In other words, should replies not be received, an 'early warning' indication has been given of failure and of its location. The only reservation here is that this is not true early warning in that in the worst case, the detection of the failure may be delayed for the duration of the system scan period.

Two design steps can be taken in this instance; the system (scan) rate itself can be speeded up and special priority can be given to critical data sources so that they are up-dated much more frequently than in the routine scan. The first step entails the use of a higher data transmission speed on the master-outstation communication link by adopting e.g. 'half-cycle' working (Section 3.2).

As far as the second of these stages is concerned, meeting the demand for the flexible software required to bring in selected sources at very short updating intervals would be extremely useful for final operation. Assuming that this repeated rapid access could be switched to any source required, it would represent a Crisis management facility held in the main plant control system.

It will be seen that a major principle is involved here - the question of determining on an 'immediate' basis, which data source has changed to becoming 'critical', and consequently should be moved on to the highest priority level. 'Alerting signals' will come up at the Master Control Position in two ways : (a) from the continuously operating - real time - Crisis management system working with critical point instrumentation established as in Section 5.2, (b) from,the main control alarm system, which, as indicated earlier, is subject to delay; and which also, with VDU page presentation, can involve protracted search and reading time.

Co-ordination of recording

It is assumed that for final operation Crisis management, i.e. both Anticipatory and Crisis control, will be based to a great extent on visual observation of continuously available 'real-time' displays. Methods are suggested in Chapter 6 which might be adopted for recording and analysing such displays; but at this stage it is taken that the fundamental principle of Crisis control applies and that such 'extraneous' complex equipment is excluded from the operational system. This means that a full printed-out tabular record would not be available as with the main system; and that events would have to be logged with reference to time markers on a graphical (chart) recorder. Such a recorder could be switched on manually when Incident or comparable conditions were being simulated, and could be supplemented by voice-recording switched on at the same time, and supplied with corresponding time reference.

Because of delay in print-out on the main system, where alarms have to be stored before being printed-out, a similar type of combined chart and voice record would almost certainly have to be used. This would, however, result in the desired common record being brought into use for the two systems.

As indicated at the beginning of this section, this work is intended to provide data for the preparation of software for the advanced schemes described in Chapter 6 should this be required. A simpler approach could be adopted where main and Crisis control systems were not so closely coordinated.

5.6   References

1.   CHARLISH, G.(1981) : 'Making sure the product will always work', Financial Times, London (Friday March 6 p.15)

2.   DEVEREUX, R.(1964) : 'Development of a long-range telemetering buoy', Undersea Technology Compass Publications Inc., Arlington, Va. USA

3.   SMELT, R.(1961) : 'The Agena Satellite and the Discoverer Programme', J. Roy. Aeron. Soc., 65, 611

4.   ADVISORY COMMITTEE ON MAJOR HAZARDS (1979) :'Second Report',
Her Majesty's Stationery Office, London
5.   HEALTH AND SAFETY EXECUTIVE (1980)   : 'Flame arresters and
explosion reliefs' Booklet HS (G) 11, Her Majesty's Stationery
Office, London

Chapter 6

TECHNIQUES FOR THE FUTURE

6.1 'Independent check' instrumentation - Possible CCTV
developments

In these proposals for new methods of instrumentation the
television techniques used to 'relay' and display the instrumentation information are, in general, more advanced than those
described for the deliberately simple schemes of Chapter 2. Thus
colour is brought in, while advantage is taken of the rugged
nature of specially designed modern TV cameras. In this last
instance, an underwater camera can be quoted as being capable of
operating down to a depth of 300 m (1000 ft) with full colour
capability.

Plate 7   Marconi V330 Radiation Tolerant Camera Head fitted in a special
camera plug assembly prior to positioning inside the shield wall of the
irradiated spent fuel dry store at CEGB Wylfa Nuclear Power Station
Supplied by Marconi Avionics Ltd

On the instrumentation side, one of the main attractions of TV, as compared with conventional transducer methods, is that the hyper-interface of the latter, as a physical contact, is eliminated. A physical/electrical conversion does, of course, take place in the camera; but this can be situated remotely from the actual instrumentation source at a limit fixed by the optical viewing distance. This ability to isolate the camera becomes increasingly important as the conditions at the (nominal) hyper-interface become more hostile; and by using glass for a window aperture which is capable of withstanding maximum temperatures of the order of $1000^{\circ}C$ the TV camera can be placed behind a separation wall which gives protection at temperatures up to this value. It seems relevant to add that a working rule for the use of 'silicon devices' in utterly hostile environments is that they should not be held continuously at a temperature of more than $150^{\circ}C$; and that some form of external cooling will be required for reliable working above this temperature.

It must be pointed out that the use of TV cameras, as proposed, is not entirely a matter of isolation from adverse environments as mentioned above, although this facility is obviously of great value and could be of considerable significance e.g. in the event of fire developing on the plant side of the separation wall.

In many ways, however, the outstanding reasons for adopting these rather more complex TV methods are associated with instrumentation 'transfer' problems at what would be the transducer hyper-interface which arise when Crisis control conditions apply; and also because of various difficulties, largely mechanical in nature, when conventional methods cannot be used.

Thus in the latter case, there are instances where transducers cannot be attached to the piece of equipment anywhere near the desired instrumentation point; an elementary example being where a pressure pick-up point or the equivalent cannot be built into a structure because of consequent local weakening. (One possible method of overcoming this problem was suggested in Section 2.2 with the concept of fitting a 'washer' transducer under a flange bolt).

In general, however, it has to be accepted that when these unique and abnormal conditions are encountered, and the requirement for instrumentation at a specific point must be met, conventional methods of transducer 'insertion' must be abandoned. As a result, alternative techniques have to be found which ideally do not require more than surface contact with the 'structure' involved.

As already indicated, the TV base does give a starting point for evolving such techniques, and certain possible approaches are introduced below.

The first of these approaches is comparatively obvious - the use of temperature-sensitive paints - which, when viewed by a colour TV camera represents a 'one-shot' warning method which might well provide key information on a sudden temperature build-up during an emergency, and which might not be obtainable in any other way.

An extension of this principle is afforded by the use of 'visual strain-gauges'. A mechanical equivalent of the compact strain-gauge, i.e. the extensometer, has already been suggested as being suitable for use with the simple TV methods of Section 2.3; but with appropriate strain detection devices it seems that it might be possible to derive pressure and temperature information from the strain 'readings' in addition to the more direct stress indications.

An advanced type of visual strain gauge which perhaps could be developed for this application is based on methods of stress analysis employing polarised lights. These methods were demonstrated publicly by Professor E.G. Coker in London in the early 1900s, with an account written by him appearing in *Engineering* at the beginning of 1911. In a contemporary book (Ref.1), where Professor Coker's work is described in broad terms, coloured plates pay tribute to the remarkable results achieved, not least in the beauty of the stress colour patterns which were obtained. These photoelastic stress patterns were produced by passing polarised light through models of engineering components made from xylonite (celluloid, pyroxylin), one of the earliest plastics - a solid solution of cellulose nitrate and camphor or other plasticiser. Application of stress to the model resulted in the appearance of coloured fringe patterns capable of interpretation not only as regards stress distribution but also for its magnitude and direction. This change in optical properties with applied stress is characteristic of many transparent materials, especially plastics; and in a modern automobile research centre (Ref.2) where these techniques are being used extensively, the plastic material is applied as a contoured coating. An ingenious method of avoiding the use of transmitted light, as in Coker's original work, is to cement the photoelastic coating to the specimen with a reflective adhesive.

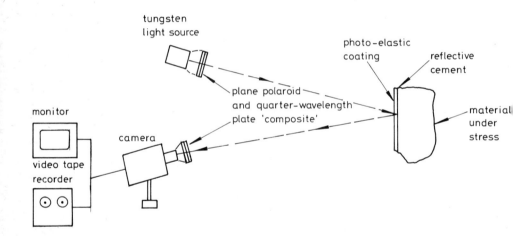

Fig. 6.1 *Basis arrangement of reflective photoelastic measurement system*
for remote - CCTV - observation

From the above it would appear possible to give at least some form of 'semi-quantitative' calibration to a strain-gauge equivalent made of this material; and then to derive from the stress patterns as seen with polarisation, the values of the parameters such as pressure as mentioned at the beginning of this account.

As far as the development of colour cameras is concerned, work appears to be directed in two main areas : the reduction in size and complexity of the cameras themselves, and improvement of the resolution performance of the solid-state 'CCD' (charge coupled device) electronic tube replacement.

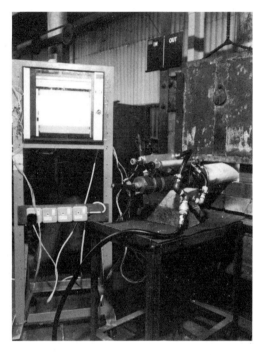

Plate 8   Prototype testing of the Marconi Furnace Viewing Probe Assembly. In the picture the camera cooling system is being tested with the main camera probe assembly actually inside the furnace at a temperature of 1300°C Supplied by Marconi Avionics Limited

An earlier major development was the production of a single-gun camera tube to replace the three pick-up tubes (for the three 'primary' colours - red, green and blue) in the original colour cameras. In terms of broadcast 625 line system, the resolution of these single-gun tubes is limited; and a figure of 410 lines has been given at the present stage of development. Such resolution is more than adequate for the applications being discussed here.

For the CCD solid-state cameras, the resolution is reduced

still further. In 1980 it was given as 226 horizontal picture
elements and 492 vertical elements on a chip approximately
10 mm X 9 mm. Two-colour cameras can be made by combining red
and blue images; and it would appear that even at this stage of
development, performance would be adequate for most of the
envisaged applications.

Finally, mention should be made of the television-associated
technique of infra-red 'thermal imaging'. This is part of the
'Detection' family which includes gas and fire/smoke warning
systems, and comparable radiation pyrometer installations.
Although these systems have a certain amount in common with the
TV based instrumentation systems as described, it is desirable
not to mix the two groups at any point, and least of all at the
monitoring position. This is particularly true in the context
of Crisis control; and it can be stated that for a large scheme,
a design study confirmed that the warning system should be
entirely independent of the main installation and should have its
own display console.

6.2 'Voice command' and control speech techniques

The following brief review is intended to put into
perspective the methods which may possibly be put forward in the
future to deal with a total Crisis Control situation i.e. when
a full 'Incident' state of emergency has developed.

It will be assumed that a multiple VDU-Console control
scheme is in operation with parallel chains for main plant and
Crisis management - including anticipatory - control. In these
circumstances there would be a master control position supplied
on a selective basis, as with the other consoles, with 'Alarm
and Situation' type information from both chains, and as
described in the next section.

With a relatively small system, direct 'real-time' control
could probably be exercised from the master control position,
with raw crisis management information being interpreted and
passed on from that console. With increase in system size,

control functions would have to be divided between the consoles which whould either be in the same control room, but separated from each other; or, in some cases, as indicated in Chapter 5, would be split away in physically separate master and sub-control rooms.

Clearly, in the last instance, there is a requirement for control speech communication to be maintained continuously between master and sub-control rooms if real-time control is to be carried out. Actually, the same requirement exists with 'common control-room' working; and, in practice, has to be satisfied by the same kind of communication method as with the separated system i.e. by some form of telephone-like link. It will be realised that this is necessary on the grounds of streaming information alone, and helping to prevent the confusion which simultaneously uttered speech could cause; but for Crisis working, as already implied, it is vital to have a permanent record available of all *relevant* control speech commands and interchange.

Taking the apparently simple aspect first, the suggestion for the communication system of basing it on a combined headset-microphone has several limitations, although it does meet, to some extent the 'eyes free, hands free' criterion applicable to such operational work. One of the main difficulties with the telephone combination set is that the wearer is cut off 'audibly' from the rest of the outside world, i.e. from those not 'on' the common (telephone) communication channel. With other than the larger schemes, this isolation is probably not a great dis-advantage and helps to reduce distraction; but with a major complex and a crisis situation, the senior engineer will require to keep in contact with the common telephone information and, at the same time, be able to keep in direct touch with other control engineers round him.

This leads to the concept of 'Two-Tier' control communication, and, by the same token, what amounts to a Two Tier Command Structure for dealing with emergencies. On the lower tier - operator - communication channel would be found information from the main control system together with Crisis management information derived from critical point and similar instrumentation sources. The last set of information - perhaps in 'filtered' form - would be fed across to the upper tier channel which would be concerned primarily with the taking of immediate decisions with regard to crisis control actions.

The question which arises in this general connection is whether the load on either communication tier - effectively a command channel - could be eased by bringing in some form of computer generated speech. Bearing in mind that only emergency conditions on a large complex are being considered, it will be clear that by the unexpected nature of such an emergency which has to be assumed, pre-programming is not possible. Thus, in view of the 'real-time' operation required, any 'keying-in' of speech would not really be practicable.

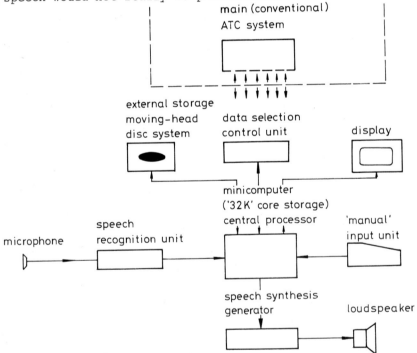

Fig. 6.2 *Speech recognition and generation system outline for Air Traffic Control training*

On the other hand, a considerable amount of work has been

done in the USA on the application of such techniques to Air Traffic Control; and at the National Aerospace Electronics Conference held in 1975, two Papers (Refs. 3 and 4) were given which bear directly on this question. These Papers are of specific interest because they cover 'automated speech' systems on which development has been completed during the formative period of these technologies. Consequently, the stage reached at that time is directly matched to the present consideration in that 'firmed-up' development would probably depart further from the emergency requirements of Crisis control as noted earlier. It should also be added that speech recognition techniques are not brought into this survey in view of the circumstances attending an emergency situation.

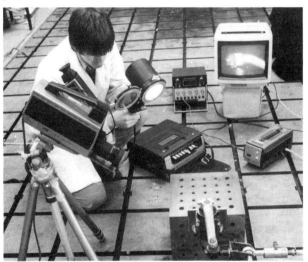

Plate 9   *Direct polariscope monitoring of television based photoelastic measurement system. Component under (hydraulic ram) stress in foreground Supplied by BL Technology Ltd.*

In the paper by Grady and Herscher, the main emphasis is placed on a GCA (Ground Controlled Approach) Controller Training System; and in the description of the various elements in the system and its operation, it is made clear that although the simulated conditions are realistic, the basic technical and operational objectives are semi-routine in nature. Nevertheless

the methods of training are of direct interest to any organis-
ation considering possible training of personnel for crisis
control operation, not least for their technical content.

The second paper by Beck and Anderson contains 14
References; and they give a valuable of techniques developed for
computer generation of speech after describing a number of
proposed systems.

From both papers it becomes clear that Air Traffic Control
represents an almost ideal field for the application of computer
generated speech. Messages are of short duration and almost
entirely routine in character, and therefore the storage required
is not prohibitively large. Thus, subject to these conditions
being realised - as in a restricted training exercise - speech
generation techniques might well be found useful for limited,
essentially routine, use in a large control complex. However,
it would seem that as soon as even a small degree of operating
flexibility was demanded in programming the cost of storage and
the complexity of the software would become excessively high.

As an illustration, a recent article (Ref.5) on 'An extremely
low-cost' voice generation system can be quoted as giving the
following statement on storage : "Sixteen spoken words can be
easily stored in 8 k bytes of memory".

6.3   'Super-software' development

From preceding chapters, and particularly from Section 5.5,
it will be clear that with any control system incorporating
Crisis management facilities, provision should be made to design-
in 'assistance' for interpreting visual displays under 'saturation' conditions.

This assistance has two basic divisions : 'Inherent' and
'External'; and its first function in both divisions is to help
to ensure that the onset, even of threatening conditions, is not
missed. In a simple Crisis management system, detection of
incipient trouble is essentially a matter of interpretation of
raw data visual displays; and with experienced control personnel

who are in continuous contact with the operation of the main
plant, 'warning signs' will almost certainly get picked up by
combining information from the two systems.

However, it becomes more difficult to be sure of catching
such signs as the complexity of the system under control
increases; and this means that some kind of alerting signal is
required. For the simple system where the raw output signals
from a 'frequency' transducer are being monitored visually as
waveforms, an abrupt change in frequency can soon be checked. In
this case, assistance can be given by establishing frequency
limits on discriminator circuits fed from the transducer output,
and arranging for audible warning to be given when these limits
are passed.

An 'inherent' type of alerting assistance can, however, be
provided in this case by taking advantage of the properties of
the analogue dial meter, covered from another standpoint in
Section 4.3. The strength of such dial presentation does seem
to lie in the 'at-a-glance' properties of the pointer ('needle')/
scale combination to which the observer's photographic memory
appears to respond, in effect, instantaneously, and with surprising accuracy. Taken in conjunction with the ability to detect
rapid changes by virtue of pointer movement, this mode of
presentation can be quite powerful as a means of giving
'alerting ' assistance.

In all but the smallest schemes, it is assumed that these
Crisis management indications would be brought together on an
Alarm and Situation Diagram. It has already been implied that
the essence of such diagrams is that they should be kept relatively simple, and therefore that the hard-wired diagram would
have to be replaced by some form of large-screen television
display when more than one instrumentation group has to be
displayed as a 'layout' picture. When the need to switch Crisis
management instrumentation sub-systems is taken into account, as
suggested in Section 3.1, it becomes clear that computer-based
switching is almost essential, and required on such a scale that
'Super-software' is a necessity, as has been noted.

Reverting to the concept of pure Crisis management philosophy, which is not met by computer-based instrumentation/display
methods, the use of colour television introduces an element of
alerting assistance in terms of change of colour; and it is
envisaged that comparatively small TV colour displays would be
brought together in an A and SD configuration - as with the
(analogue) dial and similar presentation - for visual warning
and analysis by the operator.

Finally, it becomes evident that with a large 'complex',
the quantity of the Crisis management information - not least
from Critical point instrumentation - would become so great that
complete coordination of it would be necessary. Furthermore,
this coordination process would have to be arranged to determine
the full extent of an Incident (and if possible the events which
led up to it) and the exact location of all failure points.

To meet this demand, it is suggested that a 'Crisis Overview'
system, based on TV and computer techniques, would have to be
developed to work in parallel with the 'conventional' Crisis
management display, which itself would be left untouched for
operation under full emergency conditions. This Overview system
could be conveniently integrated with the Two Tier Command
system outlined in Section 6.2, and corresponding operational
duties fixed similarly at two levels in the former.

A possible form of comprehensive Crisis Overview system
would have two functions. Assuming that basic presentation would
be by large-screen colour television, and that individual Crisis
management instrumentation 'areas' were available in A and SD
form, the first function would be to analyse these areas and to
bring the area of most danger to the notice of the senior control
engineer. This action would begin as a warning, and then con-
tinue as a presentation of the danger area held by the engineer
for a time determined by his overall assessment of the situation.

The second function would be to analyse the various TV
pictures and determine the position of - in a simple embodiment -
all the red instrumentation areas. This perhaps could be
achieved by bringing in all the area TV pictures and then viewing

them with a second camera - as in the very early 'picture
(standards) convertors' - to extract the diagram coordinates of
the red indication areas.

As a footnote, it might be possible to arrive at a simple
image analysis system by adopting sequential colour field scanning for the second camera, so that position data could be
extracted for a specific colour on a given repetitive field. It
should be pointed out that the main disadvantage of sequential
colour scanning is that colour breakup takes place on rapid
motion, e.g. with a white rotating wheel, the spokes are shown
with a full colour spectrum. In the present case, it would
appear that such speeds of motion would not be developed.

6.4   References

1.   KNOX, G.D. : 'All about Engineering', Cassell and Company
Ltd., London

2.   FT (1981) 'Stress seen before it is a problem', Financial
Times, London (Friday January 4 p.9)

3.   GRADY, M.W., and HERSCHER, M.B. (1975) : 'Advanced Speech
Technology applied to Problems of Air Traffic Control', IEEE
1975 National Aerospace Electronics, pp. 541-546

4.   BECK, A.F., and ANDERSON, D.E. (1975): 'Computer-Generated
Voice in Air Traffic Control Applications', IEEE National Aerospace Electronics, pp.547-551

5.   ANDERSON, J.E.(1981) : 'An extremely Low-Cost computer
voice Response System' BYTE, BYTE Publications, Inc.,
Peterborough, N.H.

Chapter 7

GLOSSARY

7.1 Select glossary with explanatory notes

Anticipatory Control   The second functional division of Crisis
Management (q.v.). The techniques of Anticipatory Control,
associated with an Early Warning function, are aimed at forestalling as far as possible the development of Incident
conditions.

At-a-glance working   Observation of a meter or other data display
in which the information is recorded mentally without error and
without 'reading delay'.

Crisis Control   The first division of Crisis Management (q.v.)
is an intervention action where an operator takes over-riding
control in the event of an emergency, first to contain it and
then to bring it under control.

Crisis Management   The generic term covering both Crisis Control
and Anticipatory Control which are combined within it.

Critical Point   A point in a system which - as far as instrumentation is concerned - by virtue of its position can be the source
of vital information, and which, if passed unnoticed in a fault
condition, can lead to the development of a Crisis situation.

Data Marshalling   The techniques of Data Marshalling are directed
towards identifying and extracting critical data from a multiplicity of sources, of 'streaming' it into its appropriate
information channel, and of ensuring that no vital data is lost.

Demodulation   The process by which the original modulating
signal is derived and reconstituted from the modulated carrier
wave, sometimes called 'recovery' of the original modulating
signal.

Earthquake (type) Disaster   A destructive accident of a magnitude which is equivalent to that of a natural disaster as exemplified by a severe earthquake.

Hyper-interface (transducer)   The critical transfer interface at the input to the transducer (q.v.), seen as a system, where conversion is made from the 'parameter' physical state to the corresponding electrical state.

Incident Conditions   These conditions, representing a state of emergency, are set up when a completely unforeseen - unexpected - failure develops. Such situations can, in extreme cases, reach Crisis proportions.

Independent Check   This principle is one by which a specific indication of measurement is covered by two completely separate system 'chains', each working on an entirely different physical basis from the other.

Intrinsic Safety   The technique of Intrinsic Safety is designed to give protection against the occurrence of sparks or other "thermal effect" which might cause ignition of a potentially explosive atmosphere. An intrinsically safe circuit may be defined as one in which the energy contained in any spark that may occur is kept below the value at which it would be capable of causing ignition of a given explosive atmosphere.

Raw Data   The concept of Raw Data is one in which signals or similar information are kept in their original 'derived' form and are not passed through any form of data processing before reaching their point of presentation. The original usage was in radar practice in connection with 'raw radar' signals.

'Real Time' Operation   This kind of (control) operation is one in which both the presentation of information and the taking of action consequent upon it, must be 'immediate'. For this control regime to be effective, all information must have maximum 'security'.

Transducer - instrumentation   A device which responds to a physical stimulus, such as a varying pressure, and produces an electrical data signal related to it.

Control in Hazardous Environments - INDEX

R. E. YOUNG